Large-Scale Oceanographic Experiments and Satellites

NATO ASI Series

Advanced Science Institutes Series

A series presenting the results of activities sponsored by the NATO Science Committee, which aims at the dissemination of advanced scientific and technological knowledge, with a view to strengthening links between scientific communities.

The series is published by an international board of publishers in conjunction with the NATO Scientific Affairs Division

A	Life Sciences	Plenum Publishing Corporation
B	Physics	London and New York
C	Mathematical and Physical Sciences	D. Reidel Publishing Company Dordrecht, Boston and Lancaster
D	Behavioural and Social Sciences	Martinus Nijhoff Publishers
E	Engineering and Materials Sciences	The Hague, Boston and Lancaster
F	Computer and Systems Sciences	Springer-Verlag
G	Ecological Sciences	Berlin, Heidelberg, New York and Tokyo

Series C: Mathematical and Physical Sciences Vol. 128

Large-Scale Oceanographic Experiments and Satellites

edited by

Catherine Gautier

California Space Institute A-021,
Scripps Institution of Oceanography, La Jolla, California, U.S.A.

and

Michèle Fieux

Laboratoire d'Oceanographie Physique,
Museum National d'Histoire Naturelle, Paris, France

D. Reidel Publishing Company

Dordrecht / Boston / Lancaster

Published in cooperation with NATO Scientific Affairs Division

Proceedings of the NATO Advanced Research Workshop on
Large-Scale Oceanographic Experiments and Satellites
Sainte-Lucie de Porto-Vecchio, Corsica, France
3-7 October, 1983

Library of Congress Cataloging in Publication Data

NATO Advanced Research Workshop on Large Scale Oceanographic Experiments and Satellites (1983 : Sainte-Lucie de Porto-Vecchio, Corsica)
Large scale oceanographic experiments and satellites.

(NATO ASI series. Series C, Mathematical and physical sciences; vol. 128)
"Proceedings of the NATO Advanced Research Workshop on Large Scale Oceanographic Experiments and Satellites, Sainte-Lucie de Porto-Vecchio, Corsica, October 3—7, 1983."
"Published in cooperation with NATO Scientific Affairs Division."
Includes index.
1. Astronautics in oceanography—Congresses. 2. Artificial satellites in geographical research—Congresses. I. Gautier, Catherine, 1947- . II. Fieux, Michele. III. North Atlantic Treaty Organization. Scientific Affairs Division. IV. Series: NATO ASI series. Series C, Mathematical and physical sciences ; vol. 128.
GC10.4.A8N37 1983 551.46′0072 84-8224
ISBN 90-277-1786-9

Published by D. Reidel Publishing Company
P.O. Box 17, 3300 AA Dordrecht, Holland

Sold and distributed in the U.S.A. and Canada
by Kluwer Academic Publishers,
190 Old Derby Street, Hingham, MA 02043, U.S.A.

In all other countries, sold and distributed
by Kluwer Academic Publishers Group,
P.O. Box 322, 3300 AH Dordrecht, Holland

D. Reidel Publishing Company is a member of the Kluwer Academic Publishers Group

Printed in The Netherlands.

TABLE OF CONTENTS

ACKNOWLEDGEMENTS

We gratefully acknowledge the NATO Marine Science Committee and the California Space Institute for their financial support of the Advanced Research Workshop on Large-scale Oceanographic Experiments and Satellites held in Corsica, France, October 3-7 1983.

Several persons greatly contributed to the organization of the meeting and the preparation of the final version of this volume. Our special thanks to Beth Di Julio for overseeing the whole undertaking from its beginning to its end; to Anne White for attending to the many facets of the organization and for supervising the last minute hectic arrangements at the site of the conference; to our local hosts, particularly Madame Huguette Rocaserra, and Messieurs Geronimi and Castelli for their valuable assistance with the needed equipment; to Susan Hope for managing to prepare on time all the manuscripts put together at Scripps; to Francoise Demerson-Baker for her great help in the editing, revising and preparation of this volume; and to Renee Thoby who, on short notice, kindly assisted in the preliminary stages of the manuscript preparation in Corsica. The extra contribution in time and effort of the moderators and rapporteurs of the meeting's discussions was greatly appreciated.

FOREWORD

An Advanced Research Workshop (ARW) sponsored by NATO and the California Space Institute was held in Corsica (France) October 3 to 7, 1983 to discuss the role of satellite observations in the large-scale oceanographic experiments, especially those under discussion (e.g., the World Ocean Circulation Experiment, WOCE, and the Tropical Ocean and Global Atmosphere, TOGA).

This volume is based on papers presented during that meeting, summaries of the discussions of the working groups and recommended necessary tasks to be accomplished in preparation for WOCE and TOGA.

The participants of the meeting decided that, although the collection of issues discussed in the meeting was undoubtedly incomplete, the summaries of the discussions and recommended tasks warranted being conveyed to the organizers and sponsors of WOCE and TOGA.

Although not discussed at the workshop, it was recognized that an important role of satellites is as data collection and location systems.

Some of the common conclusions of the different working groups discussions are that:

1) Studies are needed of the sensitivity of the ocean response to errors in surface parameters (wind stress, heat flux, SST etc.) in a variety of physical models. These should be one of the basis for determining the accuracy requirements in WOCE and TOGA.

2) The ultimate calibration standards in developing satellite algorithms should be direct measurements of the fluxes rather than bulk parameterizations. Additionally, efforts should be

made to obtain these calibration measurements with some degree of arial averaging in order to decrease sampling errors in the comparison with the remote measurements.

3) Studies should be carried out to determine the nature of the satellite measurements errors, in particular whether the errors are random or systematic. This should greatly help in defining the expected accuracy of the satellite measurements over the scales of importance to WOCE and TOGA objectives and delineating the applications of the data.

4) Careful consideration should be given to the importance of simultaneous flight of TOPEX-POSEIDON, ERS-1 and NROSS. Surface topography and wind stress measurements are required to understand the dynamic response of the ocean to changes in surface wind forcing. This is one of the main objectives of TOGA. Greater confidence in the satellite measurements and a more flexible approach can be achieved with the existence of simultaneous altimeter satellites in different orbits. Intercomparison and cross calibration of the instruments and the data treatment can improve confidence in the reliability of altimeters and radiometers, in particular.

Catherine Gautier

A CONCEPT OF WOCE

James C. McWilliams

National Center for Atmospheric Research
Boulder, Colorado, 80307, USA

ABSTRACT

A discussion is presented of various candidate observational elements for a World Ocean Circulation Experiment and their relative merits for determining aspects of the general circulation. In addition, speculations are offered about the dynamics of the general circulation; in particular, the importance of "stretching vorticity" (i.e., fluctuations in the static stability profile) is emphasized.

1. INTRODUCTION

The possibility of a World Ocean Circulation Experiment (WOCE) is being discussed in a variety of fora, and it is therefore of interest to consider its design. This essay is a collection of some personal opinions about relative priorities among possible techniques of observations and about dynamical concepts for a WOCE.

My definition of the General Circulation is the distribution of and physics governing currents and related water properties on the horizontal and time scales within the following range:

$$\text{MESOSCALE} \leq (L,T) \leq (\text{GLOBAL, 5- OR 10-YEAR MEAN}).$$

C. Gautier and M. Fieux (eds.), Large-Scale Oceanographic Experiments and Satellites, 1–16.

In my opinion any finer restriction of the range of scales would be physically artificial and would lead to substantially less understandable results.

Some observing techniques can span this full range of scales; satellite systems are particularly suited to this. Inevitably, most in situ systems can only look at portions of this range. Obviously, a WOCE should place more emphasis on those which can see a greater portion. Also, however, there will be important, locally concentrated measurements at "pulse points" of the General Circulation---boundary currents, water property transformation zones, etc. In general, I believe a WOCE should include a complicated mix of measurements, and its design will be correspondingly complicated.

2. THE CENTRAL ELEMENTS OF WOCE AND THEIR RELATIVE PRIORITIES

My primary premise in considering the design of a WOCE is that fundamentally it should focus on circulation (i.e., velocity), on a variety of scales, and velocity measurements are generally a better basis for inferring other quantities of dynamical interest than are other observables, although the strength of this advantage obviously varies with the quantity and scale of interest. A secondary premise is that a WOCE can be done, which would be an enormous advance on our knowledge of the General Circulation, making use of both instruments and numerical models which either now exist or are under development with a reasonable expectation of fruition within a few years. The real challenge, in my opinion, is not to determine the feasibility of a useful WOCE, but rather to select among the possibilities and find the time, energy, people, and money to apply them.

A list of the central elements of a WOCE is presented in Table 1, ordered in three priority classes. Any such list is obviously premature and overly simplistic at the present preliminary stage of planning for a WOCE, but it should be at least provocative, if not instructive, to describe the rationale behind this particular list.

Table 1

CENTRAL ELEMENTS OF WOCE

Physical quantity	Principal measurement technique(s)
surface geopotential	altimetry + tide gauges
wind forcing	scatterometry; sea level pressure + meteorological model
mid-depth velocity	deep drifters
surface layer thermo dynamics and circulation:	
*buoyancy forcing	satellites (at least for short-wave radiation); ships; surface drifters, meteorological model
*heat content	surface drifters; XBTs; (tomography?]
*surface velocity	surface drifters; acoustic
*mixing penetration depth	logs ships; drifters
water-mass volumes and geography (i.e., "mean" lapse rate, dynamic height, chemical tracers; potential vorticity)	shipboard hydrography and water sampling
variability of dynamic height, tracers, and interior heat content	shipboard surveys; tomography; purposeful tracer releases
long-path velocity and area-averaged relative vorticity	tomography with reciprocal shooting; deep drifters (heavily averaged)
+ a variety of more local measurements (as required)	

The first priority class comprises my concept of a minimal WOCE. Without all of these elements, I would be reluctant to defend the experiment as encompassing a sufficient portion of the presently unknown aspects of the General Circulation to be worth the bother of a WOCE. Of course, any of these elements would be valuable for a variety of oceanographic purposes, but less so than a WOCE. Altimetry and scatterometry have the right sampling characteristics to span the full range of General Circulation scales, as defined above. The wind forcing might also be obtained by a complementary technique, a network of sea-level pressure measurements which would be converted to wind forcing by a suitable meteorological model. Mid-depth velocity (i.e., below the thermocline) is the best complement to surface geopotential for a full-depth determination of the velocity field (see Sections 3 and 4), and deep drifters are the most likely technique for sampling at least approaching the desired range of scales.

The second priority class includes buoyancy forcing. It is often considered secondary in magnitude to wind forcing in driving the General Circulation, although this is certainly not true by the measure of potential energy in the oceans even if it is by kinetic energy. Buoyancy forcing is also more difficult to deal with than wind forcing---practically, because available techniques are not likely to be as relatively accurate for the net forcing, and conceptually, because its "transmissivity" to the General Circulation below the surface boundary layer is much more subtle than for momentum. For this latter reason, WOCE should address the surface layer as a whole rather than simply the surface buoyancy fluxes. The global water masses have historically been the most useful indicator of the largest scales of the General Circulation, and any significant gaps in the historical collection should be filled in WOCE. These quantities provide the environment for the circulation; tracer distributions provide contraints upon possible circulation patterns (we are not yet very good at spelling out these constraints, partly because sources, sinks, and boundary conditions for tracers are not very well known); mean dynamic heights shore up the altimetric measurements where they are least accurate (at zero frequency, where the geoid must be subtracted); and large-scale potential vorticity is closely related to useful dynamical concepts for the large-scale circulation (Section 5).

Variability in the water masses has been assigned to the third priority class for two reasons: it is very labor intensive to approach the desired sampling rate (although tomography can in principle do much better at this than shipboard measurements), and dynamic height is relatively inefficient, compared to the combination of surface geopotential and mid-depth velocity, in determining circulation (see Sections 3 and 4 below), except in regions where the vertical structure of currents is particularly complicated. Obviously an important practical distinction for WOCE---which I won't attempt here---is the scale boundary between more useful "mean" water properties and less useful and more difficult to measure "variability". An interesting possibility in this latter category, which might not be subject to the two disadvantages stated above, is one or many purposeful tracer releases. The path-averaged velocity measurements which might be achieved with reciprocal shooting tomography might yield the largest horizontal scale components of the deep circulation more accurately than averages of deep drifter trajectories, and a connected circuit of them might yield the area-averaged relative vorticity within the circuit. One can state several reservations about this information---we have as yet no experience interpreting such information; it is uncertain how much of the energy of the deep circulation is on very long horizontal scales; it is unclear how usefully dynamical concepts of the large-scale can be related to relative vorticity, which is the usually neglected component of the large-scale potential vorticity (see Section 5); and it is unclear how fine a horizontal resolution (i.e., smallest path length) is feasible. However, if satisfactory answers to these present uncertainties can be found, this technique may be a very valuable one for WOCE.

The final entry in Table 1, more local measurements, is important for WOCE, because certain features of the General Circulation are quite localized, but this will not be addressed further in this talk.

Finally, I would remark that the priority classes in Table 1 are not intended as sequential limits to the physical quantities and measurement techniques to be included in a WOCE. I believe all of these quantities and techniques should be included. The priority classes rather are intended to indicate the relative amounts of effort that might go into the various alternatives.

3. THE RELATIVE VALUE OF VELOCITY AND GEOPOTENTIAL MEASUREMENTS IN DETERMINING THE CIRCULATION

Part of the rationale for the ranking in Table 1 is the relative value of velocity and geopotential measurements in making estimates of the circulation. We can rely on the geostrophic relationship,

$$\underset{\sim}{v} = \hat{e}_z \times \underset{\sim}{\nabla}\phi$$

which allows us to consider their information content as equivalent in the absence of measurement or sampling errors. However, in the

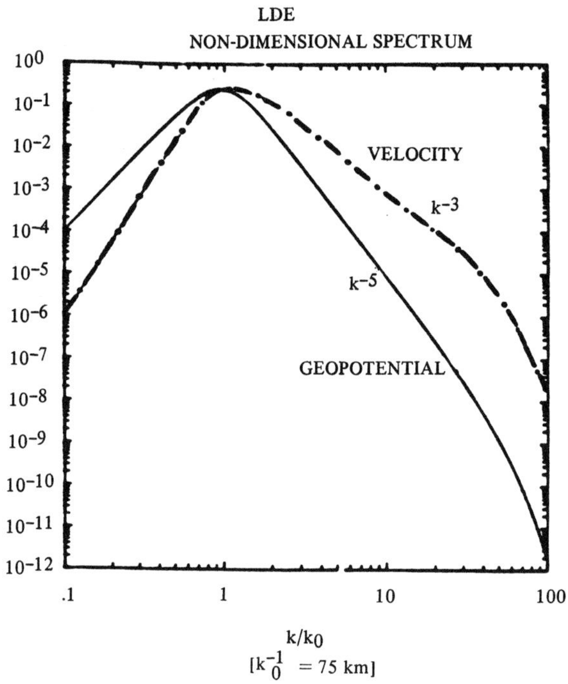

Fig. 1 (from McWilliams, Owens, and Hua, 1983)

presence of these practical deficiencies, the fact that, by geostrophy, the two fields must have different wavenumber spectra implies that geopotential measurements are relatively inefficient, compared to velocity measurements, in estimating velocity.

The differences in spectra are illustrated in Fig. 1, which is a statistical model fit to the observations in the POLYMODE Local Dynamics Experiment (LDE). Its empirical accuracy is least at the largest and smallest wavenumbers, where the spectrum shape has been chosen with theoretical prejudice, and greatest at intermediate wavenumbers. No wavenumber spectrum for the General Circulation is well known, of course, but this one from the LDE is among the best presently available.

A demonstration of the relative efficiency of the two types of measurements is the following. Consider a one-dimensional array of N data points, xi, uniformly distributed over an interval of length L. At each of these points, either geopotential or velocity measurements are available. In either case we make optimal linear estimates of the velocity at any point, x, within the interval with the formulae

$$\hat{v}(x) = \Sigma_{i,j} \langle v(x)\phi(x_i)\rangle (\langle\phi(x_i)\phi(x_j)\rangle)^{-1} \phi_j$$

or

$$\hat{v}(x) = \Sigma_{i,j} \langle v(x)v(x_i)\rangle (\langle v(x_i)v(x_j)\rangle)^{-1} v_j,$$

where the angle brackets denote ensemble mean covariances. The expected errors in such estimates are, respectively,

$$\langle(v-\hat{v})^2\rangle = \Sigma_{i,j} \langle v(x)\phi(x_i)\rangle (\langle\phi(x_i)\phi(x_j)\rangle)^{-1} \langle\phi(x_j)v(x)\rangle$$

or

$$\langle(v-\hat{v})^2\rangle = \Sigma_{i,j} \langle v(x)v(x_i)\rangle (\langle v(x_i)v(x_j)\rangle)^{-1} \langle v(x_j)v(x)\rangle .$$

The required covariances are evaluated from

$$\langle\phi(x)\phi(y)\rangle = C(x-y)[1 + \varepsilon^2\delta_{x,y}]$$

$$\langle v(x)\phi(y)\rangle = C'(x-y)$$

$$\langle v(x)v(y)\rangle = C''(x-y)[1 + \varepsilon^2\delta_{x,y}] \quad ,$$

where C(r) is the Bessel transform of the LDE geopotential spectrum (Fig.1),

$$C(r) = \int_o^\infty kJ_o(kr)S(k)dk$$

and ε^2 is the relative error variance in a point measurement, modeled as white noise. For simplicity the same value for ε^2 is taken for each of the types of measurements. A summary measure of the estimation error is defined by

$$\langle\delta v^2\rangle/\langle v^2\rangle \equiv \int_o^L \frac{dx}{L} \langle(v-\hat{v})^2\rangle/\langle v^2\rangle,$$

and this quantity is plotted in Fig. 2. The result is the anticipated one that, for a given N, errors from velocity measurements are smaller than those from geopotential measurements.

This result is reasonably robust in spectrum shape (except for very red shapes), the relative error level, and the interval length. If the relative errors are sufficiently larger for the velocity measurements, of course, their competitive advantage can be reversed, although this is unlikely to be true in prac-

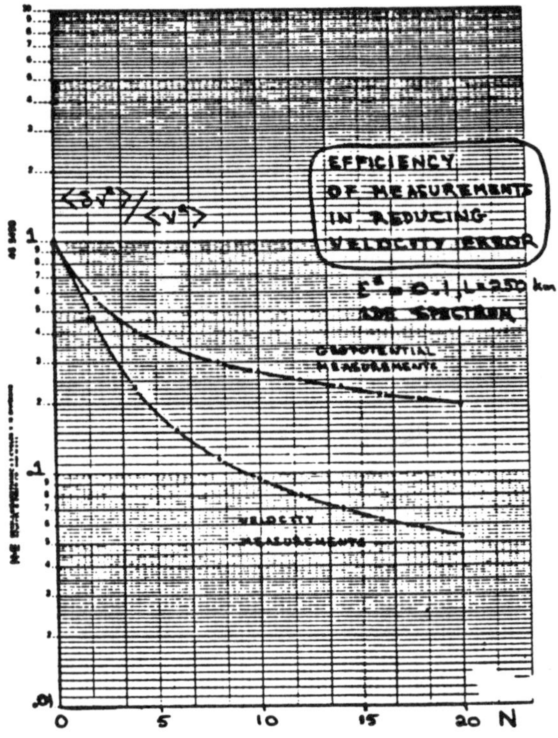

Fig. 2

tice. If alternatively we were to define the desired estimand as only the very large scale component of the velocity, then a competitive advantage would be given to geopotential measurements, opposing the tendency demonstrated above, since intermediate and small scale components of the spectrum in effect become errors in the estimation and the magnitude of this is larger for velocity than for geopotential.

In general, however, I believe the competitive advantage for WOCE will be found to most often lie with velocity measurements in estimating the circulation, and this is one of the reasons for giving dynamic height and temperature tomography measurements (both of which are closely related to geopotential) lower priority than mid-depth velocity measurements in Table 1. The same criticism, of course, can be applied to altimetric measurements of the surface geopotential, except that there is no serious alternative for obtaining surface geostrophic velocity at the desired sampling rates.

4. VERTICAL SAMPLING ISSUES

Another reason for the previous assignment of priorities is due to the distribution of currents in the vertical. Consider a truncated modal representation for geostrophic velocity of the form

$$\underset{\sim}{v}(\underset{\sim}{x},z,t)=\Sigma_{n=1}^{M}\underset{\sim}{V}_n(\underset{\sim}{x},t)f_n(z) + \delta\underset{\sim}{V}(\underset{\sim}{x},z,t),$$

where the fn(z) are the familiar modes from linear theory, the $\underset{\sim}{V}_n$ are the modal amplitudes, and $\delta\underset{\sim}{V}$ is the residual associated with truncation after M modes. We model $\delta\underset{\sim}{V}$ as uncorrelated with any of the modal amplitudes and with modal amplitudes and with itself except at zero lag. It has been shown, from long time series from moored current meters in several locations, that this representation is often an efficient one even with M = 2. Two examples of this are shown in Table 2; note that the depth-averaged relative error variances are quite small. The two sites have quite different modal statistics---the LDE has about three times the amplitude of MODE and its degree of cross-modal coupling is significantly less---associated with their different mean circulation environments (the LDE is within the Gulf Stream Recirculation Zone, while MODE has much weaker mean currents). The modes themselves vary slightly between the two locations, due to differences in the mean lapse rate, but this variation is secondary.

Table 2

MODAL STATISTICS (M = 2)

Quantity		MODE		LDE	
		m=1	2		
$\langle\underset{\sim}{V}_n \cdot \underset{\sim}{V}_m\rangle$ $[cm^2 s^{-2}]$	n=1	19		57	
	2	7	13	10	31
$\varepsilon^2 = \dfrac{\int_{-H}^{o} \frac{dz}{H} \langle\delta\underset{\sim}{V}^2\rangle}{\sum_n \langle\underset{\sim}{V}_n^2\rangle}$		.047		.054	

With these modal statistics one can examine several aspects of different sampling schemes in the vertical. A very simple question is the degree of correlation between currents at different depths. With the modal representation, we write

$$\langle \underset{\sim}{v}(z_i)\cdot\underset{\sim}{v}(z_j)\rangle=\Sigma_{n,m}\langle \underset{\sim}{V}_n\cdot\underset{\sim}{V}_m\rangle f_n(z_i f_m(z_j)[1+\varepsilon^2\delta_{z_i,z_j}] ,$$

and evaluate it for three levels: at the surface, in the thermocline, and at mid-depths. The results are shown in Table 3.

Table 3

CROSS-LEVEL VELOCITY CORRELATIONS

	MODE			LDE		
	0	-700	-2500	0	-700	-2500
z = 0m	1.00			1.00		
-700	.94	1.00		.93	1.00	
-2500	.23	.33	1.00	.23	.40	1.00

They demonstrate the general result that surface and thermocline geostrophic currents are highly correlated, hence nearly redundant measurements, while deeper currents are only weakly correlated with the shallower ones, hence a nearly independent measurement.

Another question one can pose is this: suppose one has measurements of $\underset{\sim}{v}(z=0)$ and $\underset{\sim}{V}_2$---the first baroclinic modal amplitude---how well can one estimate $\underset{\sim}{v}(z)$? This question is an idealization of the possible combination of altimetric measurements of the surface flow with either dynamic height or tomographic temperature measurements of the baroclinic flow, disregarding here the inefficiency of converting geopotential to velocity information (Section 3). It also can be answered in the context of optimal linear estimation theory using the preceding modal representation and Table 2 statistics. The results are shown in Table 4. Errors in estimating $\underset{\sim}{v}(z)$ in and above the thermocline are satisfactorily small, whereas they are quite large below the thermocline. This and the preceding calculation clearly indicate the importance of a set of deep velocity measurements.

Table 4

$\underset{\sim}{v}(z)$ ESTIMATION ERRORS FROM $\underset{\sim}{v}(0)$ and $\underset{\sim}{V}_2$

z [m]	$\langle(\hat{\underset{\sim}{v}}(z)-\underset{\sim}{v}(z))^2\rangle/\langle\underset{\sim}{v}(z)^2\rangle$	
	MODE	LDE
-700	.10	.13
-2500	.53	.46

It is unclear how generally an M = 2 expansion is an efficient representation of the General Circulation, and geographical exploration of this issue would be very useful for the design of WOCE. It seems likely that this representation will be less successful for circulation on the largest scales than it has been shown to be for the point variability sampled by one-to-several-year moorings, although we cannot discount the possibility that some other low-order modal representation, perhaps with different modes, might be comparably efficient. In any event, if, as also seems likely, the largest scales of circulation are more surface intensified than the mesoscale and intermediate scale currents which dominate the MODE and LDE statistics, then the necessity for deep velocity measurements will be even greater for adequate estimates of deep circulation.

Thus the arguments of both the preceding and present Sections tend to lower the priority accorded to the baroclinic geopotential in the context of altimetry and deep velocity. Some of the counter-arguments are listed in Section 2.

5. LARGE-SCALE DYNAMICS

At the risk of both oversimplification and as yet far too little empirical justification, I propose that the point dynamics of the large-scale, low-frequency (i.e., beyond mesoscale) General Circulation are, to a substantial degree, the advective-diffusive dynamics of stretching vorticity.

Potential vorticity can be split into three components for small Rossby number:

$$q = P + S + R,$$

where P represents the combined effects of Coriolis frequency variations and topography, S the vertical stretching of fluid columns, and R the relative vorticity. The first two of these are often expressed as f/h, where f is the Coriolis frequency and h is the thickness of a fluid layer bounded by surfaces of constant potential density or the ocean bottom. In this notation the mean potential vorticity equation is

$$\frac{\overline{D}}{Dt}\left(\overline{P} + \overline{S}\right) = -\nabla\cdot\overline{\underset{\sim}{v}'S'} \quad + \quad \begin{array}{l}\text{wind forcing}\\ \text{buoyancy forcing}\\ \text{relative vorticity}\\ \text{sub-mesoscale effects}\\ \ldots\end{array}$$

Here the overbar denotes large-scale, the first term on the right is associated with mesoscale motions, and the list refers to other processes generally of lesser magnitude. This equation expresses what I mean by the dynamics of stretching vorticity. It is much more likely to be valid as a point balance, which is the form in which it is written, than in a large area or volume integral, where the "lesser" terms will rise in relative importance.

The accompanying heat equation is

$$\frac{\overline{D}}{Dt}\,\overline{T} \approx \overline{w}\,\frac{\partial\overline{T}}{\partial z} = -\underset{\sim}{\nabla}\cdot\,\overline{\underset{\sim}{v}'T'} + \ldots$$

(advection) (diffusion)

This balance has been demonstrated at the LDE site in the Gulf Stream Recirculation Zone, where both the mesoscale eddies and large-scale circulation are strong, and may carry-over to other, weaker regions. The inferred vertical velocity is plotted in Fig. 3. An indication of the dominance of stretching vorticity dynamics is the relative smallness of the Subtropical Gyre Ekman pumping (wE) at the upper surface. Estimates of the strength of mesoscale diffusion at several locations are shown in Fig. 4. It clearly has significant geographical variation, and in some locations is quite large. There appears to be a monotonic increase with mesoscale eddy energy, although there is no reason to believe that the straight-line fit has general validity.

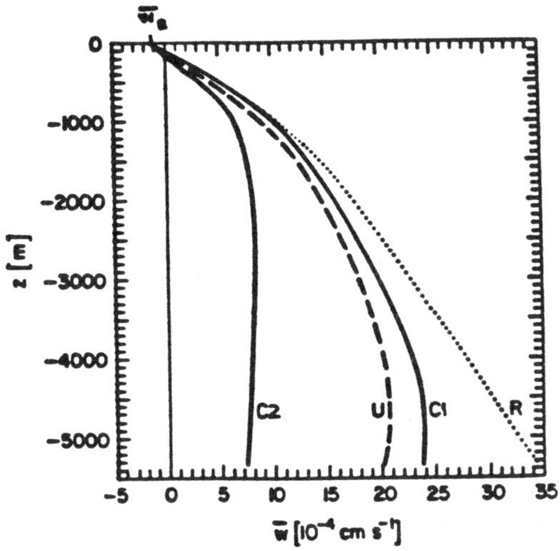

Fig. 3. Several inferential estimates of mean vertical velocity at the LDE site (McWilliams, 1983).

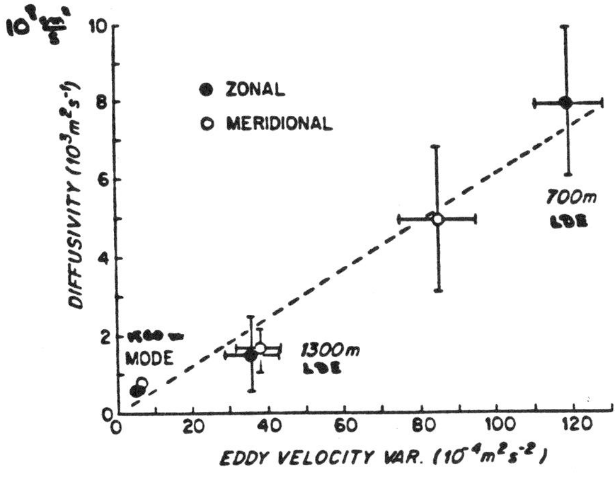

Fig. 4. Horizontal, single-particle diffusivity vs. the variance of eddy velocity (Price, 1983).

Several studies have recently focused upon the large-scale distribution of f/h. A particular interest in these studies has been the possibility of spatially extensive, homogeneous regions where this quantity has little variation. For several years now, eddy-resolving numerical-model solutions have exhibited homogeneous regions when the interior, sub-mesoscale diffusivities are sufficiently small. An illustration of f/h in the North Atlantic is shown in Fig. 5. In a thin layer in the upper thermocline, there is a region which might be homogeneous;

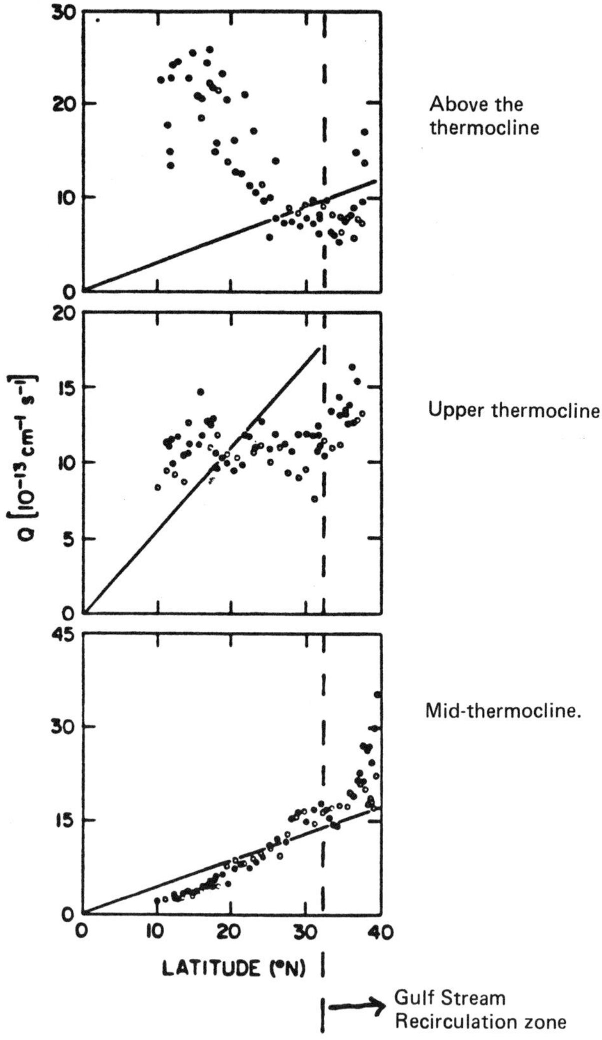

Fig. 5. P + S values along 65° and 50°W. Longitude (solid and open circles, respectively) for three layers bounded by different σ_θ surfaces (McDowell, Keffer, and Rhines, 1983). The line in each panel indicates the contribution from f variations alone.

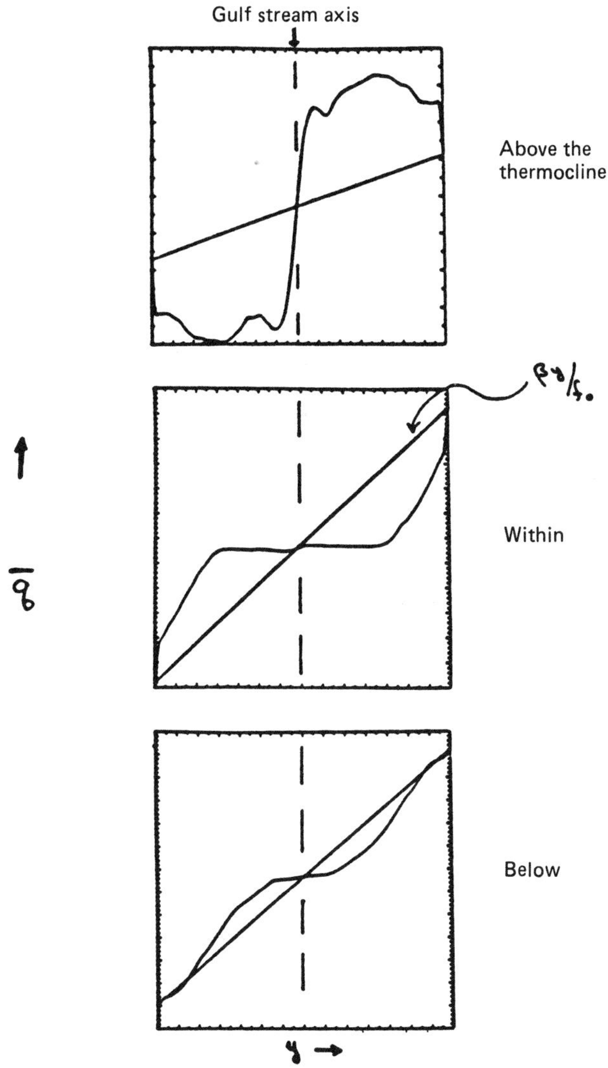

Fig. 6. $\overline{P} + \overline{S} + \overline{R}$ profiles from a numerical model solution (Holland, personal communication).

certainly any systemmatic variations are weaker than the contribution from f alone. There is substantial scatter about a possible uniform value which, if not either measurement noise or compensated by local relative vorticity, would suggest that the large-scale f/h is not exactly uniform (n. b., in a truly homogeneous region both the mesoscale and the large-scale q are uniform). In any event, the possibly homogeneous region ends at the outer boundary of the Gulf Stream Recirculation Zone.

An example of $\bar{q}$ profiles from a two-gyre, numerical model solution is shown in Fig. 6 (in most locations one can neglect the difference between q and f/h). In this case there are extensive homogeneous regions, extending through the Gulf-Stream and its Recirculation Zones. This solution appears to be much more homogeneous than the North Atlantic. However, the North Pacific appears to be much more homogeneous than the North Atlantic (Keffer and Rhines, personal communication) and may therefore be better matched by this model solution.

I certainly hope that a WOCE will yield good empirical descriptions of the elements of stretching vorticity dynamics and the distribution of potential vorticity on the larger scales of the General Circulation, so that theoretical ideas on these topics can be better assessed.

6. DISCLAIMERS

There are many aspects of a WOCE which I have not talked about and are fully as important as those I have---the reasons for wanting to do a WOCE, special regions, numerical modeling, tracer and water mass transformation dynamics, human and institutional arrangements needed for accomplishing a WOCE---and this neglect is only for lack of time.

REFERENCE

McDowell, S., P. Rhines and T. Keffer, 1983, North Atlantic potential vorticity and its relation to the general circulation. _J. Phys. Oceanography_, 12, 1417-1436.

TOGA OVERVIEW

F. Webster
University of Delaware, USA

M. Fieux
Laboratoire d'Océanographie Physique, France

ABSTRACT

The international climate research program entitled the Interannual Variability of the Tropical Ocean and the Global Atmosphere (TOGA) will study year-to-year climate variations in the tropical oceans and the global atmosphere, and the mechanisms which determine those variations and their predictability. In this paper we review the scientific objectives of TOGA and present a summary of TOGA data needs, with an emphasis on satellite sensing.

1. INTRODUCTION

This overview summarizes the international climate research program entitles the Interannual Variability of the Tropical Ocean and the Global Atmosphere (TOGA). For the purposes of this Workshop, we stress the oceanic aspects of the linked atmospheric and oceanic climate phenomena and of the proposed research program. Furthermore, we emphasize the needs of the TOGA research program for ocean observations by satellite-borne sensors.

The aims of TOGA are to determine the nature of the interannual variability of the tropical oceans and global atmosphere and to understand the mechanisms which determine that variability and the predictability of the variations.

2. THE SOUTHERN OSCILLATION

Interannual climate fluctuations have been known and

C. Gautier and M. Fieux (eds.), Large-Scale Oceanographic Experiments and Satellites, 17–24.

studied for many decades. Though the TOGA program is intended to deal with the global tropical ocean and the entire atmosphere, most attention has been centered on the near-global phenomenon known as the "Southern Oscillation". The Southern Oscillation is marked by a large-scale seesaw movement of mass in the atmosphere between the Eastern and Western hemispheres, and is centered over the eastern Pacific Ocean, with planetary dimensions. An index of the Southern Oscillation, the sea-level pressure at Darwin, Australia, is correlated with sea-level pressure over a significant part of the globe and with precipitation over much of the tropics and lower mid latitudes. The occurrence of the Southern Oscillation is irregular, but tends to happen at two-to-seven-year intervals. Though it is generally characterized as an atmospheric mass exchange, the Southern Oscillation in fact is a family of linked oceanic and atmospheric climate phenomena. Among these are year-to-year variations in the monsoon over India, droughts over Brazil, flooding in coastal Peru and Ecuador, and climate anomalies over North America and elsewhere.

The Southern Oscillation is associated with persistent sea-surface temperature (SST) anomalies. SST anomalies occur in all tropical oceans, but are particularly well developed in the eastern tropical Pacific Ocean. They correspond to a large heat content and provide a memory for atmospheric climate fluctuations over a period of many months. Though correlations have been found between the SST anomalies and precipitation patterns (heavy rainfall and droughts, for instance) the mechanisms of connection are not yet understood and their study will form a major part of the TOGA research program. The warming of the sea surface in the eastern tropical Pacific is known as the El Nino phenomenon. It has profound damaging effects on the fisheries off the coast of South America, and can cause disastrous coastal flooding in Ecuador and northern Peru as a result of excessive rainfall.

3. OTHER INTERANNUAL CLIMATE PHENOMENA

SST anomalies in oceans other than the Pacific have been shown to be correlated with rainfall in other parts of the tropics. For instance, droughts over northeast Brazil correlate with SST anomalies in the South equatorial Atlantic, Indian monsoon rainfall is correlated with Arabian Sea SST anomalies. Australian rainfall is correlated with eastern Indian Ocean and Pacific Ocean SST patterns.

Tropical SST is also correlated with interannual variations in extratropical climate conditions, particularly in the winter hemisphere. Anomalously warm SST in the equatorial Pacific is associated with mild winters over northwestern North America and severe winters over the southeastern United States. By analogy, there is evidence of a relationship between tropical Atlantic SST anomalies and wintertime conditions over northwestern Europe.

The time scale of interannual climate phenomena such as the Southern Oscillation and its linked climate phenomena is long enough so that there is promise of some climate predictive ability as each occurrence of the process evolves. The correlations mentioned above between equatorial sea-surface temperatures and subsequent climate anomalies are only rudimentary at this time. However, such correlations hold promise that models of global climate variability might be developed and used for prediction of climate anomalies a season in advance. Before this can be realized, however, it is likely that the physical mechanisms of the interannual variability will have to be described and understood. This is the objective of TOGA. To reach that objective, a program of observations, numerical modelling, and theoretical studies will be required.

4. SCIENTIFIC ISSUES

The scientific issues, as we see them today, should guide the establishment of TOGA program priorities and of the observational requirements. These issues are of three types:

1. The atmospheric response to SST anomalies.

2. The response of the tropical ocean to sea-surface wind stress and to air-sea energy exchanges on time scales of months to seasons.

3. The mechanisms of the coupled ocean-atmosphere system.

These issues for TOGA will here be summarized in turn.

Sea-surface temperature patterns and the associated current systems in the upper equatorial oceans appear to provide the memory for global interannual climate fluctations. The distribution of precipitation in the tropics appears to be sensitive to the distribution of SST, particularly over the warmer parts of the tropics, such as the western Pacific ocean. Experiments with numerical models of the general circulation give results that are generally consistent with the observed correlations between tropical SST and atmospheric climate anomalies.

For TOGA, we seek to understand if and how large-scale atmospheric circulations arise in response to SST anomalies. The atmospheric response will result from vertical fluxes of heat and moisture from the ocean to the atmosphere. To study both the fluxes and the consequent large-scale atmospheric response, there will be a high-priority need in TOGA for satellite monitoring of tropical SST anomalies and of the sea-surface fluxes of heat and moisture on a global scale. Because of the limited areas of the tropical ocean that have been studied by conventional means, satellites will be particularly important for collecting measurements in those regions (such as the Indian Ocean) where data are sparse.

Those phenomena in the Pacific Ocean that are related to the Southern Oscillation are the only responses of the tropical ocean to atmospheric forcing that have been extensively documented. The SST anomalies have been relatively well studied. SST in the eastern tropical Pacific shows the greatest interannual variance. Changes in sea level in the tropical Pacific are also linked with the Southern Oscillation. The surface changes in sea level can be related to sub-surface oceanic structure. During a warm event in the eastern tropical Pacific, sea level rises along the coast of the Americas and falls in the western Pacific.

In the Atlantic, the oceanic response is less well described, but there is some evidence for an El Nino-like equatorial SST anomaly. In the Indian Ocean, where the data are sparse, the interannual variations of the meteorological parameters are better known than the oceanic ones. Recent results have shown that the monsoon wind system and the Pacific trade-wind field are strongly coupled at the interannual time scales and that the decrease of the trade winds is associated with a warming trend in SST over the entire ocean (as in 1972). On the other hand, an increase in the trades is associated with cooling. SST and air temperature anomalies seem to be related to the Southern Oscillation phenomenon, but extensive observations are needed. Furthermore, recent results from modelling studies have shown that the atmospheric circulation is more sensitive to SST in the warmer parts of the ocean, and the Indian Ocean is among the warmest areas of the Tropical Ocean.

Though the documentation is meager, there is some evidence that ocean currents patterns respond to interannual climate variations.

The primary atmospheric driving force on the tropical oceans is wind stress. There is evidence for this from the relationship observed between wind stress and sea level in the western Pacific. Furthermore, models have shown that the observed wind stress over the central and western Pacific can produce sea-level responses in the eastern Pacific that match observations.

TABLE 1

SATELLITE OCEAN DATA REQUIREMENTS FOR TOGA
(Excerpted from the TOGA Study Group Report)

Data type	Spatial Resolution (lat/long grid)	Temporal Resolution	Accuracy
Surface temperature	2°	Monthly av.	$0.3^{\circ}C$
Surface salinity	2°	Monthly av.	$\pm 0.1^{\circ}/oo$
Sea level	2°	5-day av.	± 2cm
Surface current	$1\text{-}2^{\circ}$	monthly av.	± 0.1m/s
Wind	2°	twice daily	± 2m/s (instantaneous) ± 0.2m/s (average)
Surface Flux Measurements:			
Radiation	2°	twice daily	
Moisture flux	2° (with 1° in some regions, e.g., the equator)	monthly av.	$\pm 5W/m^2$
Heat flux		monthly av.	$\pm 15W/m^2$
Wind stress		twice daily	
Precipitation	2°	daily total	3cm/month in average

5. OBSERVATIONAL NEEDS

Ocean monitoring for TOGA should be focussed on processes with time scales of weeks to years. Of particular interest is the climatological mean annual cycle. Many of the phenomena of interest have large spatial scales. This poses a particular problem of coverage and highlights the importance of satellite sensors, to be addressed at this Workshop. The Workshop should consider how satellites might meet TOGA's needs for ocean monitoring.

One step might be to review the existing defined TOGA needs for satellite observations. Such needs were set out by the Tokyo Workshop on Large-Scale Experiments in the WCRP (World Climate Research Program, 1983). These were reviewed and refined by the TOGA Study Group (1983), whose report presents a table of TOGA data requirements. The ocean data needs presented in that table are here summarized as Table 1.

The TOGA Study Group report points out that since inter-annual variability is the primary concern of TOGA, monthly mean values will likely be sufficient for most needs. However, though the Study Group defined the need for monthly means, a higher frequency of observation and archiving was recommended. This will provide the means to resolve the interannual variability of the statistics of the fluctuations. In many cases the data are already collected at a higher frequency. This should be continued, thereby giving individual investigators the means to examine processes, and providing a data set for such things as model validation.

For TOGA, most of the oceanic features of interest are large-scale, and the spatial resolution generally recommended in Table 1 is 2° of latitude and longitude. The value of 2° is given consistently in the Table in order to encourage standardization in data resolution. It is further recommended that the grid boxes be centered on even latitudes. Note that the recommended resolution does not resolve fine-scale features such as the Intertropical Convergence Zone and the equatorial undercurrent.

Though at first sight Table 1 seems to provide reassurance that the satellite data requirements have been established for TOGA, in fact the requirements list is of limited value. It may even be counter-productive. For many of the variables, a single specification may not be realistic. It is often necessary to know the purposes for which the data are needed before a meaningful specification of requirements can be made. For example, the need for wind data may really be a need for wind stress data. The assumption made in Table 1 is that wind stress can be calculated from wind. But is this the best way to determine wind stress?

What perhaps is needed, and which this Workshop might provide, is a more analytical definition of data needs, based perhaps on the process to be studied. For each variable, the resolution and accuracy should be defined for each of the processes to be studied. Such a list will look messier, and may give the impression of less certainty than now appears to be the case. Nevertheless, it will be more valuable in taking the next steps in developing satellite systems to support TOGA.

6. TOGA DATA NEEDS

For TOGA, the next critical data needs are:

Surface heat flux The latent heat flux is difficult to measure, since its estimation depends on relative humidity, surface wind stress, surface air and sea temperature. Precipitation is related to latent heat release in the atmosphere. Most of these variables are difficult to measure on the scale required by TOGA. On the other hand, the apparent critical importance of surface heat fluxes in interannual climate variability makes this topic the highest priority.

Surface wind stress The wind stress, as the driving force exerted on the ocean by the atmosphere, is particularly important. Existing methods of measurement, such as cloud winds and ship reports, need review to establish their comparability. Algorithms to relate them may need to be developed. Satellite-borne scatterometers hold the most promise for surface wind-stress estimates over the entire tropical ocean.

Surface currents Away for the equator, sea-surface topography from satellite altimetry has the potential to provide the field of surface flow. At the equator, however, currents are important because of their role in heat advection. However, the currents there are not geostrophic and the sea-surface slope cannot provide the means to estimate the currents.

Sea-surface temperature Existing accuracies from satellite radiometers and from ships seem adequate for defining the large-scale patterns of SST and their anomalies. Somewhat better accuracy will be needed for future studies of processes.

CONCLUSIONS

The most immediate need in TOGA is for an adequate description of the processes of interannual variability in the ocean and atmosphere. The need is specially keen in the ocean, where even a rudimentary description is lacking for some regions, such as the eastern Indian Ocean.

When the descriptive base becomes stronger, the next step will be to formulate coupled models of the ocean and global atmosphere. These in turn will be of value in refining the observational network so as to minimize the forecasting errors in the models. As the data base builds we will have an improved ability to validate our models.

The outcome of this Workshop will be of value to TOGA if it defines, for each process to be studied, the critical measurements that may be collected by satellite. For those measurements where our existing capability is inadequate, it will be most helpful to TOGA to highlight the need to develop better means for observation.

REFERENCES

Climate Research Committee, 1983, El Nino and the Southern Oscillation. National Academy Press, Washington, DC, 72 pp.

Committee on Climatic Changes and the Ocean, 1983, CCCO Panel on Indian Ocean Climate Studies, First Session, 21-23 March 1983, Paris, Intergovernmental Oceanographic Commission, Paris, 9 pp. + 17 app.

TOGA Study Group, 1983, Report of the JSC/CCCO Study Group on Interannual Variability of the Tropical Oceans and Global Atmosphere (TOGA). WCP-49 World Meteorological Organization, Geneva, 55 pp.

World Climate Research Program, 1983, Large-Scale Oceanographic Experiments in the World Climate Research Programme, WCRP Publications Series, No. 1, World Meteorological Organization, Geneva, Vol. I, 121 pp.; Vol. II, 544 pp.

POTENTIAL FUTURE ALTIMETRY MISSION: A JOINT NASA-CNES EFFORT

J.L. Fellous
CNES, France

W.S. Wilson
NASA, USA

Discussions between NASA and CNES in 1981-1982 have resulted in a decision to conduct a joint definition study examining the possibility of cooperation between the proposed NASA TOPEX (1) and CNES POSEIDON (2) missions.

The objective of a potential combined mission is to measure the surface topography of the global oceans with sufficient accuracy and precision such that the measurements may be used in determining their general circulation. An additional objective is to prepare for follow-on oceanographic missions allowing long-term observation of the slowly varying large-scale ocean currents.

The scope of this joint definition study was agreed upon by NASA Administrator, James Beggs and CNES President, Hubert Curien in a meeting in Paris on June 7, 1983. It was agreed that NASA and CNES would study the feasibility of a cooperative program based upon the following elements:

- NASA would provide the TOPEX spacecraft, consisting of an appropriate bus and carrying the full TOPEX sensor complement.

- CNES would provide the ARIANE launch vehicle for the TOPEX spacecraft to be placed into the planned TOPEX orbit; the DORIS tracking system; and, if possible, a solid state amplifier for the TOPEX altimeter or a separate POSEIDON altimeter sharing the TOPEX altimeter antenna, depending on the results of the joint definition study.

C. Gautier and M. Fieux (eds.), Large-Scale Oceanographic Experiments and Satellites, 25–26.

- U.S. and French scientific communities would cooperate in the definition of the science plans for the mission through appropriate joint working groups and scientific committees. In particular, these groups would define and propose plans for data processing calibration for the scientific exploitation of the mission. Joint scientific activities would be based on the assumption of equal access of both scientific communities to the data collected during the operational phase under conditions proposed by the joint working groups and consistent with established data availability guidelines.

This joint definition study was initiated in the Summer of 1983 and will extend until September, 1984. The earliest launch date would be in early 1989, subject to NASA FY-86 new start approval and to a CNES program decision by mid-1985.

In addition to this joint definition study, both NASA and CNES are conducting independent studies. NASA is studying the launch of TOPEX aboard the space shuttle and CNES is studying the launch of POSEIDON aboard the SPOT-3 satellite. Though no commitment is made for the development of either project or of a follow-on joint activity by this definition study agreement, there is a strong feeling that such a cooperative effort could play a key role in contributing to scientific programs such as the World Ocean Circulation Experiment.

REFERENCES

"Satellite altimetric measurements of the ocean: report of the TOPEX Science Working Group." 1981, Jet Propulsion Laboratory, Report 400-111, 78.

POSEIDON: Objectifs Scientifiques, CNES, 1983.

A SAMPLING STRATEGY FOR ALTIMETER MEASUREMENTS OF THE GLOBAL STATISTICS OF MESOSCALE EDDIES

Francis P. Bretherton

National Center for Atmospheric Research
Boulder, Colorado, 80307, USA

ABSTRACT

Estimation of mesoscale eddy statistics is but one objective of satellite missions involving altimetry, and but one of many important applications of the altimeter. The discussion which follows is intended to contribute to a common understanding of the value and feasibility of this objective, and to focus on requirements and desiderata which may be in common or in conflict with other applications, particularly concerning orbital repeat strategy. The principal conclusions are:

Repeat tracks to an accuracy of 10 km every 3-5 days are needed for an observing period of at least 6 months, and preferably longer, together with simultaneous measurements of sea-state. The short repeat period is to reduce contamination by ionospheric and atmospheric moisture fluctuations. Useful eddy statistics at lower frequencies could be obtained from randomly drifting or long period repeat orbits during the remainder of the mission.

1. Why mesoscale eddies

The realization during the past 20 years that the ocean circulation is fundamentally turbulent has been a major advance in our qualitative understanding. Quasi-horizontal eddy velocities with length scales of 50-100 km and time scales of a month or so dominate the kinetic energy in most regions of the ocean, and are believed to be an essential part of the processes maintaining the structure of the steady current systems, as well

C. Gautier and M. Fieux (eds.), Large-Scale Oceanographic Experiments and Satellites, 27–40.

as determining the lateral mixing. Intensive case studies such as MODE and POLYMODE have established that the local dynamics can be reasonably well modelled by the quasi-geostrophic equations, and the vertical eddy profiles are fairly simple. Surveys using current meters, SOFAR floats and satellite tracked surface drifters have shown that eddies with an apparently similar structure but widely varying intensity occur all over the world ocean. Eddy-resolving numerical models of the wind-driven circulation in the North Atlantic are beginning to show considerable quantitative agreement with observations, and indicate the importance of eddies induced by barotropic and baroclinic instabilities in limiting the large scale currents and density gradients. Indeed, the eddy intensity is a sensitive indicator of the basin-scale adiabatic dynamics.

Of course, other processes such as the mean abyssal flow and mixing across isopycnal surfaces are also important to a complete picture of the general circulation of the world ocean, and to understanding its impact on issues such as climatic changes induced by burning fossil fuel. However, knowledge of the intensity and space-time structure of the mesoscale eddies as a function of position over the ocean basins of the globe would be an important constraint even on semi-qualitative reconstructions, and would be a critical cross-check on the dynamically consistent circulation models which are now under development. Satellite altimetry is well suited to obtaining this information, and naturally provides the global coverage that is extremely difficult if not impracticable from in situ instruments such as drifters. SEASAT provided a tantalizing glimpse of what can be done, but its untimely failure left the basic need untouched.

2. The mesoscale frequency-wave number bands

The intensity of mesoscale eddies may be measured by the variance in sea-surface elevation in a few selected wavenumber-frequency bands. Detailed maps of elevation are not required. For the purpose of this analysis these bands will be defined to extend from wavelengths of 150 km to 1500 km and from periods of 18 days to 180 days, with a low frequency extension beyond 180 days. For dynamically significant space and time scales these limits should be divided by 6. A breakdown of the variance to provide a low-resolution wavenumber-frequency spectrum would be important confirmatory evidence of the local statistical structure of the eddies, and indeed is essential for unambiguous interpretation. Area and time averages of these local spectra should be obtained as a function of geographic position, with a spatial resolution which is discussed below. Though satellite remote sensing can only provide information at the ocean surface, reasonable extrapolation about subsurface intensities can be made from in situ case studies such as the POLYMODE Local Dynamics Experiment.

Near the equator, east-west propagating disturbances of considerable interest fall within this frequency-wavenumber range. However, the motivation and tradeoffs of equatorial studies are distinctly different from those of mid-latitudes and they will not be discussed further here.

3. The expected signal

The estimates in this section are naturally tentative, intended only to serve as a benchmark against which to judge various tradeoffs in experimental design.

On the scales of interest velocities are in geostrophic balance, so along-track gradients in sea level equal the transverse component of velocity times the Coriolis parameter divided by gravity. The time-variance of near-surface velocities is shown for the world ocean in figure 1, estimated from ship drifts by Wyrtki (1976). It is believed that most of this is associated with mesoscale fluctuations, with a predominant length scale somewhat greater than Rossby radius of deformation, which is about 40 km in mid-latitudes and elsewhere inversely proportional to the Coriolis parameter. RMS mesoscale height fluctuations thus tend to vary with the velocity fluctuations, times a factor about 0.3s, independent of latitude. The plotted kinetic energy per unit mass is half the total velocity variance, and for reasonably isotropic eddies is equal to the variance of the transverse component. Thus the 400 $(cm/s)^2$ contour should correspond to an RMS sea-level fluctuation of about 6 cm, and transverse velocity of 20 cm/s, centered around a wavelength of about 250 km in mid latitudes. On shorter wavelengths the height spectrum is presumably steeply red, whereas on longer wavelengths it is more nearly white, depending on how peaked the velocity spectrum really is. Figure 2 compares the breakdown by frequency in an energetic and in a quieter region, obtained from subsurface current meters. Note that only 10%-20% of the variance is in periods shorter than 30 days and one third to one half is in periods longer than 180 days.

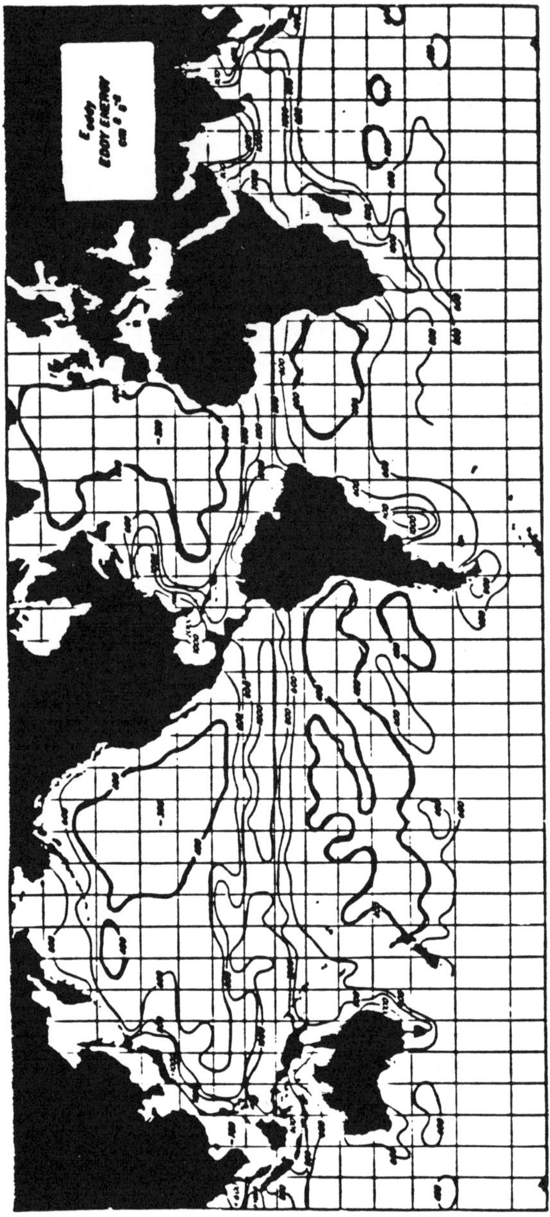

Figure 1. Eddy kinetic energy per unit mass for the world oceans based on 5° square averages. (Wyrtki et al., 1976)

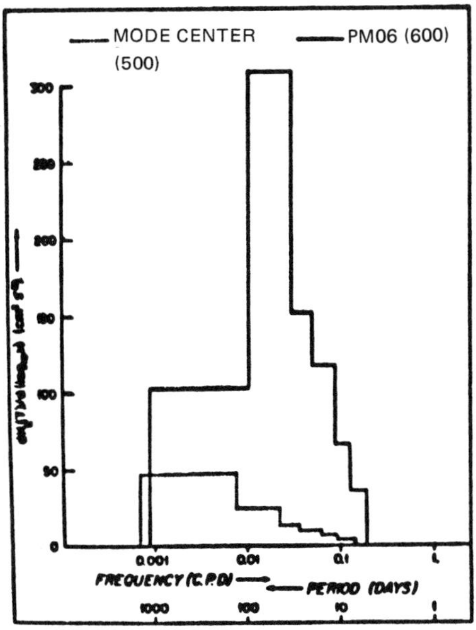

Figure 2. Estimates of the decadal frequency (ν) spectrum for the total $K_S[K\nu(T)]$ in the thermocline at MODE CENTER and at the side near the Gulf Stream (PM06; 37.5N, 55W), from depths (m) in parentheses. (Schmitz, 1978).

Based on an analysis of surface drifters in the N. Atlantic Richardson (1982) has concluded that the minimum kinetic energy of about 400 $(cm/s)^2$ reported by Wyrtki over large areas of the word ocean is probably too high, due to noise contributed by navigational errors or ship windage, and a better estimate for the median value, exceeded in half the total area would be 250 $(cm/s)^2$.

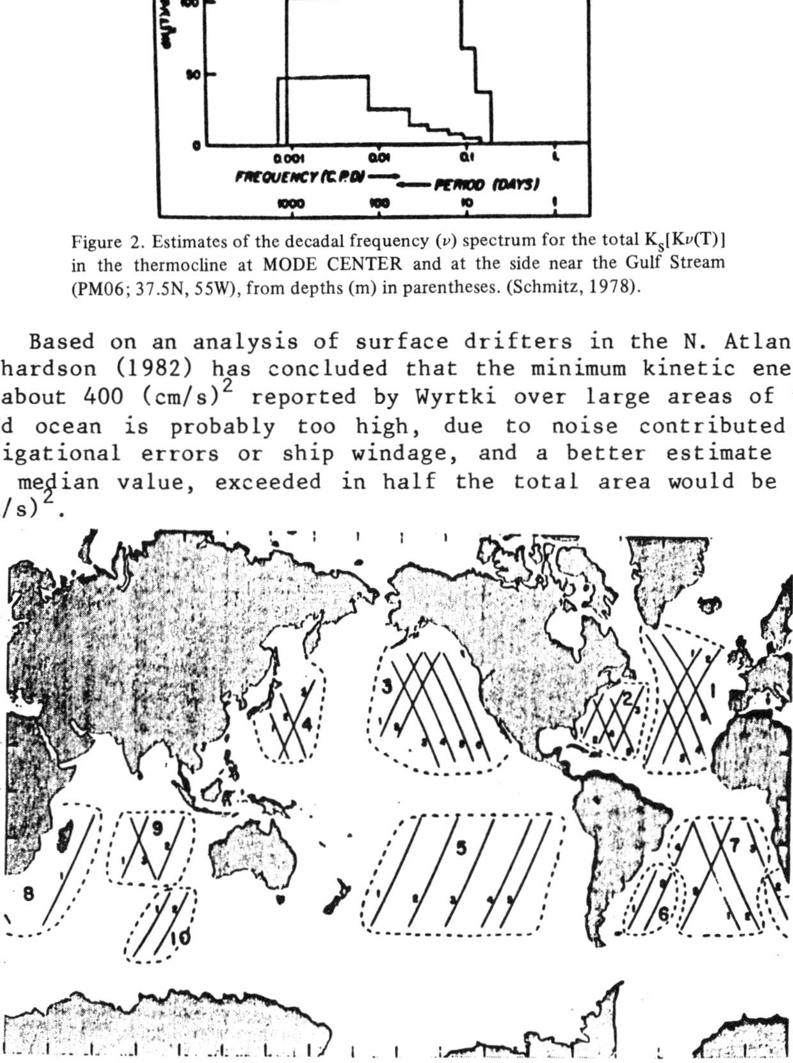

Figure 3. (Fu, 1983)

From SEASAT, Fu (1983) recently obtained along track wavenumber spectra, averaged over the areas shown in Figure 3, including all the legitimate variability of sea-level during a period of about 24 days. Figure 4 is a composite of the areas which are more energetic, corresponding to RMS height of 9 cm and cross track velocity of 30 cm/s, with a clear dominant scale for the velocity at a wavelength of 200 km. In Figure 4b are the quieter areas, RMS 2.5 cm and 10 cm/s respectively, with a blue velocity spectrum to 100 km. The ranking of the more and less energetic areas corresponds well to Figure 1, and the dominant scale in the energetic areas is just the Rossby radius of deformation. Even the variance agrees well, which in view of the short observing period indicates a substantial discrepancy with in situ experience. The internal consistency of the spectra appears good, though the different spectral shape in the quieter regions was not anticipated. This analysis is a good illustration of what can be done. Unfortunately, the SEASAT lifetime was too short to capture the major frequency range of interest.

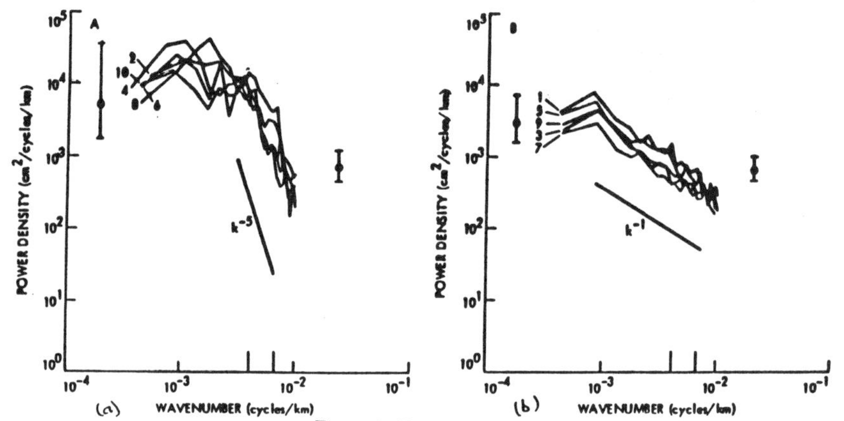

Figure 4. (Fu, 1983)

4. Analysis procedures

Most studies of sea-level variability (e.g. Cheney and Marsh (1981), Robinson et al (1983)) have been based on differences at the crossover points between ascending and descending orbits. However as pointed out by Fu (1983) the use of crossovers is not necessary to obtain mesoscale spectra. Provided a means such as repeat orbits is available to subtract mean sea-level to appropriate accuracy, long-track variations may be converted directly into spectral estimates, and statistical stability may be obtained by compositing repeated measurements along the same track and/or different tracks in the same region.

Because all tracks are in an approximately north-south direction, this yields 1-dimensional spectra, with no information about east-west variability. Because it makes fuller use of the available data, this approach will be used as the basis of the estimates which follow.

Presumably, once mean sea-level has been subtracted, meaningful differences at east-west mesoscale separations could be obtained from comparison of nearly synoptic ascending and descending orbits near but not at crossovers, but more study is needed of the information yield from different orbital patterns.

However, further study is also required of analysis schemes appropriate to uncontrolled orbits. To obtain frequency spectra it is necessary that each location be revisited to an accuracy small compared to an eddy length scale (typically 50 km), within an interval small compared to the eddy timescale. For a randomly drifting orbit 50 km track separations recur on the average every 60 days, or every 30 days if both ascending and descending tracks are included. With ingenuity, this clearly should permit a frequency analysis of periods longer than about 6 months, up to the duration of observations, but only an unresolved estimate of variability on shorter timescales. Furthermore, as remarked below, in a two year mission the mean-sea level is also adequately sampled. Thus the low frequency extension does not require repeat tracks. However, to obtain the complete frequency spectrum a planned repeat strategy for at least 6 months is essential.

5. Sources of Error

Random altimeter noise.Tracking. It is important to evaluate potential sources of error in the light of the particular wavenumber-frequency window of concern here. For example, random electronic noise and orbital tracking are easily seen to be relatively unimportant, because of the small (<100 km) and very large (>10,000 km) wavelengths on which they occur.

Ionosphere (vertically integrated electron density. More serious is the correction for vertically integrated electron density in the ionosphere. Of total magnitude about 20 cm, this can apparently be modelled as a function of time of day and latitude to an accuracy of about 30%, using as inputs the incident radiation and solar activity. The space-time spectrum of the residual is uncertain, but there is some evidence that on wavelengths of from 300 km-1500 km uncertainties increase from 1 cm up to about 3 cm. These numbers are large enough to give rise to concern in the less energetic regions of the ocean. However, given the nature of the physical processes giving rise to these small scale fluctuations it is very unlikely that they are geographically fixed on timescales of longer than a few hours, so different satellite passes are statistically independent, and repeat sampling can be of some assistance.

Atmospheric moisture (vertically integrated water vapor). In a similar category is the correction for tropospheric water vapor, for which 10 kg/m^2 column integrated precipitable water translates into 6.4 cm in sea-surface elevation. Typical corrections would be 12 cm in high latitudes and range up to 40 cm in the tropics. Fluctuations of 10% on the mesoscale would be normal. Routinely available meteorological data has inadequate spatial resolution to be very useful. On SEASAT simultaneous microwave measurements of precipitable water were available with a footprint of about 50 km, and detailed corrections were feasible in certain regions and apparently satisfactory. They show a wavenumber spectrum from individual passes which was generally small compared to the mesoscale eddy signal (Fu 1983) at wavelengths shorter than 500 km, but comparable or larger at larger scales.

Atmospheric pressure loading. Contrary to views sometimes expressed, (TOPEX Science Working Group Report 1981) atmospheric loading does not seem to be a major problem. On all wavelengths less than 10,000 km and periods of longer than 12 hrs, changes in atmospheric pressure are almost completely compensated by changes in sea-surface elevation, such as to leave the subsurface pressure gradients unchanged. This has two effects, to add noise at atmospheric frequencies to the instantaneous sea-level profiles, and to require a correction to the measured low frequency profiles to yield the mesoscale eddy velocities. 1 cm in elevation corresponds to 1mb in atmospheric pressure, or equivalently, 1 cm/s oceanic velocity corresponds to 10 m/s geostrophic wind. Though the RMS uncertainty in daily surface pressure over the oceans may indeed be as large as 3 mb, this does not seem to be a significant source of error on either count. On a single pass under an undocumented weather system, the pressure gradient might be uncertain by the equivalent of 10 m/s in the wind, but that translates into a noise of only 1 cm/s on a target signal of at least 10 cm/s. Likewise, after averaging over 1 month total fluctuations in pressure from the annual cycle are only 5-10 mb in mid-latitudes, and less in the tropics. Furthermore they are large scale (wavelengths 5,000-10,000 km), with a very red spectrum on smaller scales. Thus, for the mesoscale eddies the systematic correction is probably less than 1 cm.

Sea-state bias. The impact on sea state on estimates of the mean sea level over the footprint depends on the detailed response function of the altimeter. According to the latest estimates for SEASAT (Born et al 1982) there is a systematic bias of 7% of the significant wave height, with a residual uncertainty of 3%. With another altimeter, these numbers might be quite different. Since significant wave heights of 2 - 5 m are commonplace, and substantial mesoscale structure is to be expected, particularly in the residual, this effect cannot be neglected. Presumably the sea-state is to a first approximation independent of the mesoscale eddies themselves. Provided along track measurements of sea state are routinely collected, it would seem that the best strategy is to estimate the correction post hoc from the entire data set by correlating changes in sea-level with sea state. If necessary data from extreme cases could be ignored in the mesoscale analysis. However, it is worth noting that the 3% uncertainty was estimated from the RMS residual after regression over a limited sample of repeat passes of the changes in sea level against simultaneous measurements of significant wave height. This residual is precisely the signal which is interpreted by Fu as mesoscale eddies! It is thus not surprising that they are of comparable magnitude. However, this points up the fundamental problem of distinguishing random variations in wave bias from true mesoscale variability. A clue may rest in the space time spectra of the apparent signal. In the energetic regions the sea level variations are undoubtedly real. However, in the quiet regions the coherence time is apparently only one repeat interval (three days) or less (Fu, private communication), suggesting an atmospheric (wave bias) origin. However the spatial spectra do not look like those of variations of wave height, suggesting an oceanic origin. The evidence is thus at present inconclusive.

Tidal aliasing. Mid-ocean and earth tides are generally large scale, and can be modelled with reasonable confidence. In addition, for a sun-synchronous orbit the solar semi-diurnal and diurnal tides are aliassed into mean sea level. For the present problem this is an advantage, and the tidal correction is here not considered to be a major source of error.

Residual geoid. The geoid (or more correctly the mean sea level) enters this problem because it must be subtracted from each pass to yield the transient fluctuations which make up the eddies. Mesoscale and other variations in the geoid are so large that it has been argued (TOPEX Science Working Group 1981) that passes must be repeated to within 1 km if spatial variations are not to be aliased significantly into temporal ones. Viewing the mission as a whole, this position would appear to be unduly pessimistic, and the real requirement is for a repeat track accuracy of about 5 km. The reasoning behind this somewhat more relaxed position is as follows. The standard analysis procedure is to subtract from each member of an ensemble of repeat passes along the same nominal track the mean over the ensemble, to yield the time dependent variations. If the passes do not repeat exactly, the unknown cross track variations of the geoid (or mean sea level) are misinterpreted as time dependence. One possible approach to this problem is to estimate by regression at each point along the track (after smoothing to mesoscales) both the local mean and the local cross track slope. The actual deviation of each pass from the target track is, of course, known after the fact to high accuracy. For a reasonably large ensemble, the reduction by one degree of freedom in the spectral estimates is not serious, though there is some systematic loss of power. The most significant limitation appears to be the implicit assumption that the geoid can be represented in the cross track direction by a linear trend. The range of deviations over which this is useful depends on the geoid curvature. Except under extreme circumstances, it appears that the residual using a linear fit at a deviation of 5 km is comparable to or less than that of a deviation of a km using only the mean.

Sample size. For discussion of tradeoffs, consider an orbit which repeats every 12 days for 180 days. This samples continuously along tracks which are approximately 2.5 degrees of longitude apart, discretely in time. Unfortunately, since neighboring tracks are partially correlated, standard formulae for confidence limits on 1-dimensional spectra are inapplicable. The following is a rough replacement.

For definiteness, divide wavenumber and frequency space into octaves, defining for each the central wavelength and period as sqrt(2) times the lower limit. Then for each wavenumber-frequency band the fractional standard deviation from the true spectral estimate obtainable from an infinite sample is sqrt(2/N), where N is the number of degrees of freedom, adjusted by aggregating over an area of space and over time. The size of the space-time volume which contributes 1 degree of freedom is the irreducible grain of the analysis. For a Gaussian field, this is a long track distance equal to 0.7 times the wavelength or a time interval equal to the larger of 0.7 times the period and the data spacing. The problem is in the cross-track grain. Since we are compiling 1-dimensional spectra it is necessary to integrate, implicitly or explicity, over crosstrack wavenumber. For the spectral estimate itself this is equivalent to looking only along the track, but for the degrees of freedom the spectral shape becomes important, because the total estimate is the weighted sum of different bands each with different uncertainties. Ignoring completely problems of compositing ascending and descending orbits, a rule of thumb for reasonably isotropic situations is: If the spectrum is red, as in figure 4b, then the cross track grain is the larger of 0.7 times the long track wavelength and the cross track spacing. For a white spectrum, as the longwave portion of 4a, or far a blue one, replace the long track wavelength by the dominant band contributing most variance, in this case 250 km. Finally, note that the shortest useful long-track wavelength is set by the signal becoming comparable to random instrument noise, and for SEASAT was 100 km. The longest wavelength is arbitrary but ultimately is set by the credibility of regarding the ocean as statistically homogeneous. For observations spaced equally in time the shortest distinguishable central period is 2.8 times the spacing, i.e., 34 days, which effectively includes variations on all shorter periods. The longest distinguishable period is 0.7 times the duration of the observations, i.e., 4 months.

To illustrate these rules, consider two prototypical problems, first distinguishing with 95% confidence whether the power of the observed wavenumber spectrum differs from a standard, and second estimating to 25% the velocity variance on the scale of the Rossby radius of deformation. For the shortest period band of 34 days and less, 7.5 grain sizes are available during the 6 month record. Thus the height variance for the 1,000 km wavelength band can be determined to 50% (8 df) independently within geographic areas comprising 700 km squares, whereas for the shortest wavelengths many more degrees of freedom are available within the same squares and the uncertainty is negligible. Over 3 octaves in wavenumber, this yields an RMS uncertainty in slope of the logarithmic spectral plot of about 0.16. Thus in each square a -5/3 law could just be distinguished from a -2. On the other hand, for the frequency band centered on 4 months, only 2 df are available from time averaging, and the same degree of statistical stability requires space aggregation into 20 degree squares or equivalent area, which is acceptable in the broad ocean basins, but inadequate in the more energetically active regions near western boundary currents. This sampling situation is comparable to that for 24 day periods from SEASAT. The only remedy is a longer observation period or compositing similar regions in the manner of Fu. Note that if the repeat period were shortened to 3 days, the track spacing would be 4 times greater, but the conclusion for this problem would be much the same, because on the long wavelengths of most concern both spatial dimensions are completely sampled, though only marginally for the shorter repeat. However, contamination of the spectrum by atmospheric fluctuations is reduced by a factor of 4. Since the signal and noise spectra are additive, and the single pass noise level is comparable to the signal in many locations, the possibility of a spurious result is greatly reduced. The second problem, determining the velocity variance on wavelengths of 250 km to 25% requires 32 degress of freedom. This implies for the shortest periods geographic aggregation to about 350 km by 500 km, and about 1000 km by 1400 km for the longest. Since even with a 12 day repeat, tracks are separated by a wavelength, shortening the repeat period to 3 days to reduce contamination would have to be traded against aggregation to regions of 4 times the area.

6. Mission assumptions

For the purposes of elucidating tradeoffs in experimental design, using the altimeter aboard ERS-1, it will be assumed that

The mission will last approximately 2 years
The orbit will be sunsychronous
Repeat tracks with a range of repeat periods are possible
The altimeter will have capabilities comparable to that on SEASAT

7. Conclusions and Issues for discussion

Need for frequency spectra. The conflicts between the existing estimates of eddy intensity underscore the importance of frequency-wavenumber spectra, from periods of less than 1 month to as long as possible. This requires a repeating orbit for at least 6 months.

Development of analysis procdures. Any discussion of experimental design assumes, implicitly or explicitly, some framework for data analysis. Fu's approach to 1-dimensional along track spectra seems to yield more useful information than others so far proposed, and after extension to include frequency resolution and de-aliasing cross-track variations in mean-sea level, would appear to be the preferred technique. The approach for eliminating the geoid is new, and needs to be examined carefully. It is critical for eliminating the requirement for a repeat accuracy of 1km. Fu's approach does, however, emphasize mesoscale variations with latitude as opposed to longitude, and this sacrifices some interesting scientific information. Much can probably be recovered from relating nearby tracks, but research is necessary to quantify this. Also, it appears plausible that most of the required spectral information could be obtained if repeat tracks were available on a short repetition rate for only six months, and thereafter an uncontrolled orbital drift or very long repetition rate prevailed. However, the detailed procedures and associated uncertainties have not been worked out.

Dominant sources of error. From the discussion in section 5 it appears that the dominant sources of error in determining mesoscale spectra are likely to be sea-state bias, followed by variations in atmospheric water vapor, and ionospheric fluctuations. All three errors have a red wavenumber spectrum which could be difficult to distinguish from the desired signal, and would appear added to it. The second would be much reduced if simultaneous measurements of vertically integrated precipitable water could be routinely obtained from other instruments aboard the spacecraft. Together, they probably exceed the mesoscale eddy signal in the range of periods less than 1 month.

Since on these space scales atmospheric phenomena are essentially statistically independent on different days, a strategy for error reduction would be to repeat tracks several times more frequently than required for sampling the ocean, and time filtering. For example a repetition frequency of 3 days as opposed to 12 days would reduce the spectral noise due to atmospheric moisture and ionospheric fluctuations by a factor of 4, and would make some reduction in the impact of sea-state bias, at the price of a comparable proportionate change spatial resolution in the velocity variance maps.

Ancillary measurements. The above discussion assumes that the uncertainties in mesoscale fluctuations in sea level due to sea state bias can be reduced to a tolerable level, by estimating (post hoc if necessary) a single functional relationship between the local bias and locally measured parameters. Thus routine ancillary measurements of sea-state are essential. Likewise, local estimates of precipitable water are highly desirable.

Orbit tradeoffs. Since the operational strategy for orbit repetition has major impact on other mission objectives it must be discussed thoroughly at an earlier date. From the point of view of mesoscale spectra, repeats to an accuracy of about 10 km no less often than 12 days continued for at least six months are essential if the mesoscale frequency range is to be covered. If simultaneous measurements of precipitable water are not available it would appear to be very important to repeat more often (about every 3 days). For a 6-month observing time in repeat mode, the tradeoffs between reduced error and increased statistical uncertainty or coarser spatial resolution of the estimates are troublesome. In the judgement of the author, it is probably still preferable to have less certain results extending meaningfully over most of the ocean, i.e., for the repeat interval to be as short as possible. Thus although, if ncessary, information on lower frequency mesoscale fluctuations could be obtained for the remainder of the mission from a drifting or long repeat orbits, it is highly desirable to that the total observing time in repeat mode be as large as possible.

In situ observations. With any new observational technique, some important questions can only be resolved by intercomparison with other approaches. In particular the overall confidence in the results is vastly increased by agreement within understood limits with in situ measurements. A program of surface drifters, SOFAR floats, and perhaps current meters and hydrography, in selected regions is both feasible and necessary as an adjunct to the satellite based experiment. The adequacy of existing efforts and the initiation of additional ones for this purpose needs to be urgently considered.

Data management. The requirements for data management for this objective would appear to be similar to those for other research aspects of the mission, i.e., timely reduction and availability of both altimeter and ancillary data, careful documentation of just what algorithms and editing procedures have been used, and judicious involvement at all stages of scientists interested in the results.

Acknowledgements

The author is very grateful to Lee-Lueng Fu and to W. Brechner Owens for considerable assistance and constructive criticism during the preparation of this paper.

References

Richardson, 1982, private communication.

Robinson, A. R., Huang, N. E., Leitao, C. D., and Parra, C. G., 1983, "A study of the variability of ocean currents in the northwestern Atlantic using satellite altimetry". J. Phys. Oceanog., in press.

Schmitz, W., 1978, "Observations of the vertical distribution of low frequency kinetic energy in the Western North Atlantic", J. Marine Res., 36, pp. 298-305.

Topex Science Working Group, 1981, "Satellite altimetric measurements of the ocean", JPL Doc. 400-111, Jet Propulsion Laboratory, Pasadena, CA

Wyrtki, K., Magaard, L. and Hager, J., 1976, "Eddy energy in the Oceans", J. Geophys. Res., 81, pp. 2641-2646.

EDDY KINETIC ENERGY DISTRIBUTION IN THE SOUTHERN OCEAN FROM SEASAT ALTIMETER AND FGGE DRIFTING BUOYS

N. Daniault
Centre Océanologique de Bretagne
Brest, France

Y. Menard
Groupe de Recherche de Géodésie Spatiale
Toulouse, France

J. Gonella
Laboratoire d'Océanographie Physique
Paris, France

ABSTRACT

The eddy kinetic energy distribution in the Southern Ocean has been estimated from:

- Seasat altimeter data on repetitive orbit (24 days in September-October 1978),

- and 150 drifting buoys trajectories during FGGE (one year experiment: November 1978 to December 1979).

A good spatial coherence is found by these two independent methods in the Southern Ocean on a $5^o \times 5^o$ square grid from 30^oS to 60^oS. Nevertheless estimates of eddy kinetic energy from FGGE drifters (mean velocity averaged over 1 day) are four times larger than Seasat altimeter estimates. The distribution shows strong eddy activity near the western boundaries (Brasil, Agulhas current,...) and near topographic features (ridges and continental shelves: Drake Passage, Kerguelen Plateau, Mackarie ridge...).

The high ratio between these two estimates is explained by the fact that drifters give the total surface current (geostrophic - Ekman). This ratio tends to equal one when the same intercomparison is made with drifting buoy velocities

C. Gautier and M. Fieux (eds.), Large-Scale Oceanographic Experiments and Satellites, 41–56.

averaged over 9 days. Strong coherence between wind stress and surface current occurs mostly in the frequency band ranging from 1 cycle/2 days to 1 cycle/8-10 days with an averaged phase of 20° (current is on the left of the wind stress direction).

In conclusion, drifters still remain the best technique available to estimate the surface current field over large area where wind stress is important, mainly in the Southern Ocean.

1. INTRODUCTION

Some new insights of our knowledge of the Southern Ocean came these last five years from satellite technologies. From Seasat altimetry, Cheney, Marsh, and Beckley (1983) show in their map of global mesoscale sea height variability a nearly continuous and energetic band all around the Antarctic Circumpolar Current with isolated maxima coinciding with major topographic ridges and plateaus. During the First GARP Global Experiment (FGGE) from November 1978 to December 1979, the satellite tracking system ARGOS was used for the positioning of more than 300 buoys; about 50% of these drifters were launched in the Southern Ocean (J.F. Garrett, 1980). Because atmospheric pressure measurements were collected from these buoys, the best wind field on the Southern Ocean was made available by the European Center for Medium Range Weather Forecast (ECMRWF) in Reading (U.K.). These sets of data are used to analyse and to compare the distribution, estimated by these two independent methods, of the eddy kinetic energy in the Southern Ocean. An interpretation of the discrepancy between the estimates (altimetry and drifters) is given by the study of the coherence between wind and buoy drift.

2. SEASAT ALTIMETRY DATA ANALYSIS

Altimetry Data Collected by Seasat from September 13 to October 10 in 1978 were processed through the algorithm defined by Y. Menard (1983) to compute the averaged spectrum of the geostrophic eddy kinetic energy on the 86 collinear tracks covering the Southern Ocean from 30°S to 70°S (Fig. 1). Because of the noise in altimetry data (FU, 1983), a low pass filter was used over 50 km with resampling every 7 km (1 sec). The one dimensional wavelength spectrum of the kinetic energy for the 515 passes shows a clear peak at 220 km (Fig. 2). Divided by 2, this wavelength gives a scale of 35 km similar to a typical first mode Rossby radius of deformation in the oceans. A grid with 5° in latitude and 5° in longitude was chosen to set up the geographical distribution of the eddy kinetic energy; these dimensions correspond approximatively to twice the one dimension wavelength scale. In agreement with previous studies (Cheney et al. 1983; Lutjeharms and Baker, 1980) strong energy

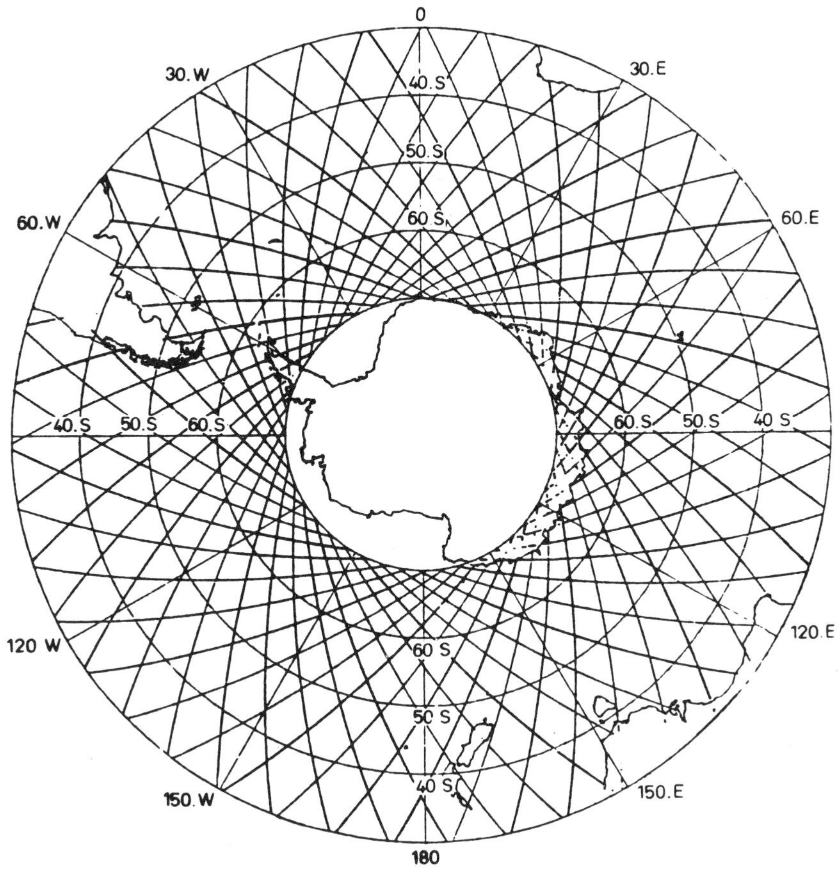

Fig. 1 Spatial coverage of the Seasat tracks from September 13 to October 10, 1978

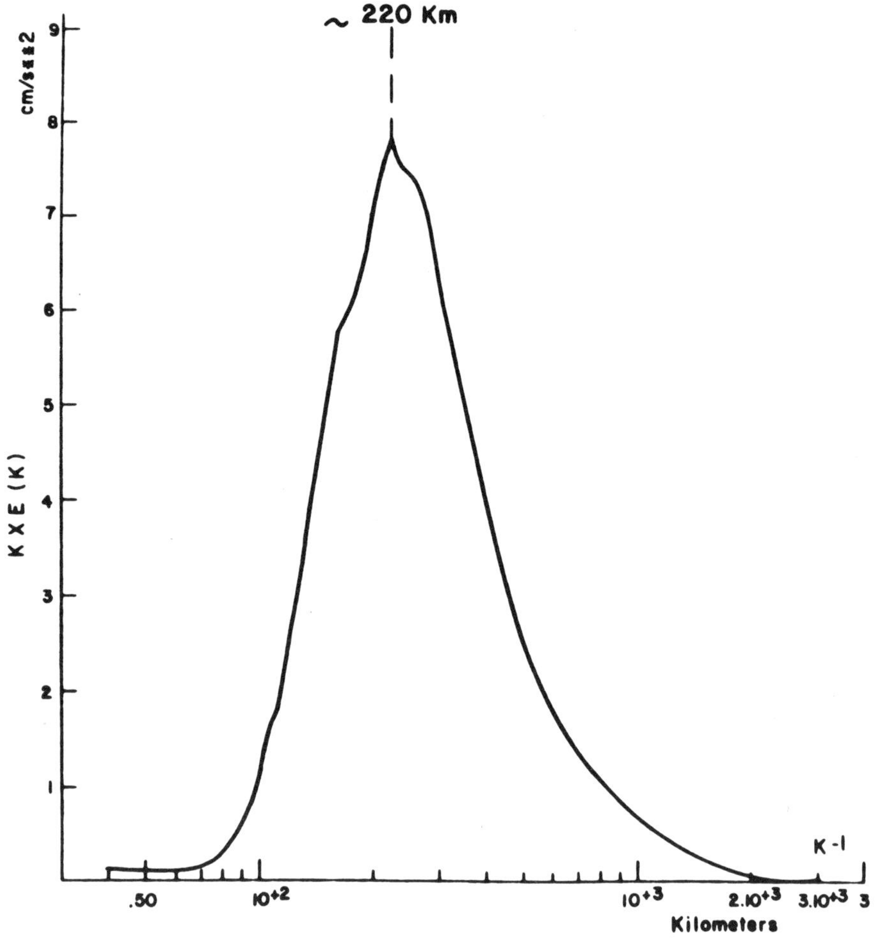

Fig. 2 Mean spectrum of geostrophic kinetic energy

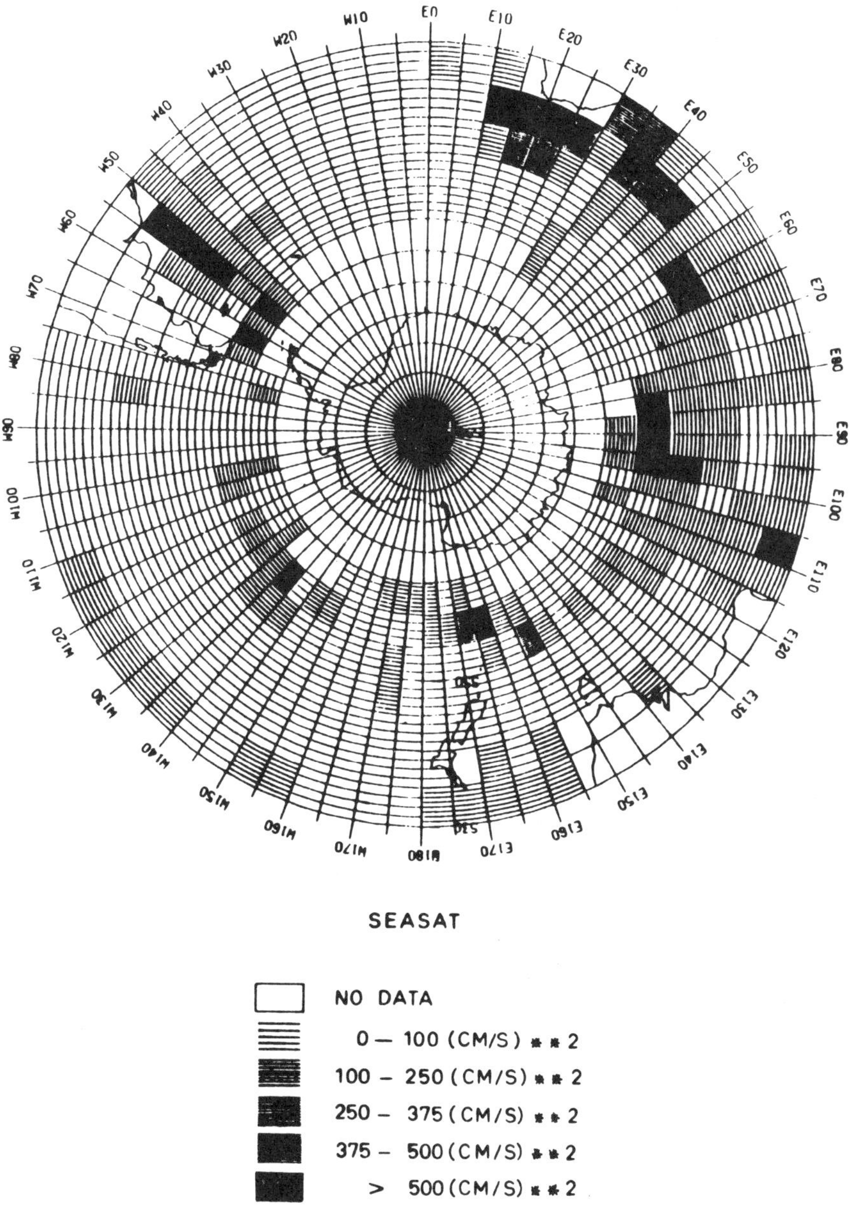

Fig. 3 Eddy kinetic energy from Seasat altimetry by $5^o x 5^o$ cells

variability (Fig. 3) is found near the Western boundaries (Brasil, Agulhas current...) and near topographic features (ridges and continental shelves: Drake Passage, Kerguelen Plateau, Mackarie Ridge...). The maximum value goes up to more than 500 cm^2/s^2 in strong energetic areas, six to seven times larger than the overall average: 80 cm^2/s^2.

3. FGGE BUOY DRIFT DATA ANALYSIS

The French Meteorological office using all the FGGE buoy data in the Southern Ocean (Fig. 4) made available the data file resampling the position every 6 hours by a cubic spline function interpolation (N. Daniault, 1982). Velocities along the paths were computed and the mean kinetic energy spectrum from 47 drifter velocity series of 300 days minimum life-time was deduced (Fig. 5). In order to remove initial strong signal, tides and noise which represent one third of the total kinetic energy variability, a low pass filter was applied on previous series with a 24 h resampling. These velocity values, for the 154 drifters used in this analysis were grouped in $5^o \times 5^o$ boxes for all the Southern Ocean. The eddy kinetic energy was calculated in the same way as Richardson (1983) did in the Northern Atlantic. For each box, containing more than 20 vector values, the departure u' (W-E direction) and v' (S-N direction) from the mean (u, v) allow to determine the Eddy Kinetic Energy.

$$E.K.E. = \quad u'^2 + v'^2 \quad / 2$$

The geographical distribution (Fig. 6) gives the same major features revealed by the Seasat data (Fig. 3) with the maximum values over 2.000 cm^2/s^2 six times larger than the mean (320 cm^2/s^2).

4. DISCUSSION

The general patterns of the kinetic energy variability determined either by Seasat altimetry (Fig. 3) or by FGGE buoy drifts (Fig. 6) are very similar but with a significant discrepancy in the estimation of the energy itself. Mean eddy kinetic energy from buoy drifts (320 cm^2/s^2) is four times larger than the mean value deduced from the geostrophic approximation from Seasat altimetry on collinear tracks (80 cm^2/s^2).

Because no statistical differences were demonstrated between drift velocities of buoy drogued in the surface mixed layer and drifting buoys with drogues that had become detached (Kirwan et. al., 1978; Mc Nally, 1981), we can expect that the windage on the FGGE buoys is not dominant, except for long term

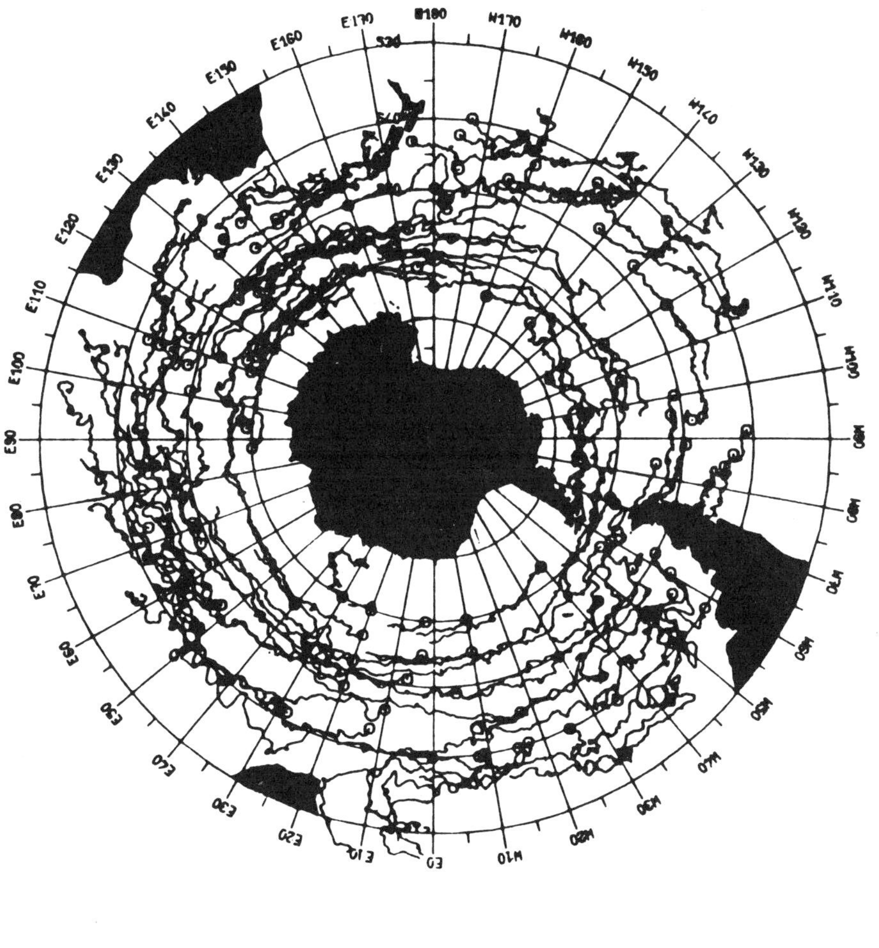

Fig. 4 Trajectories of the FGGE free drifting buoys, from December 1978 to November 1979

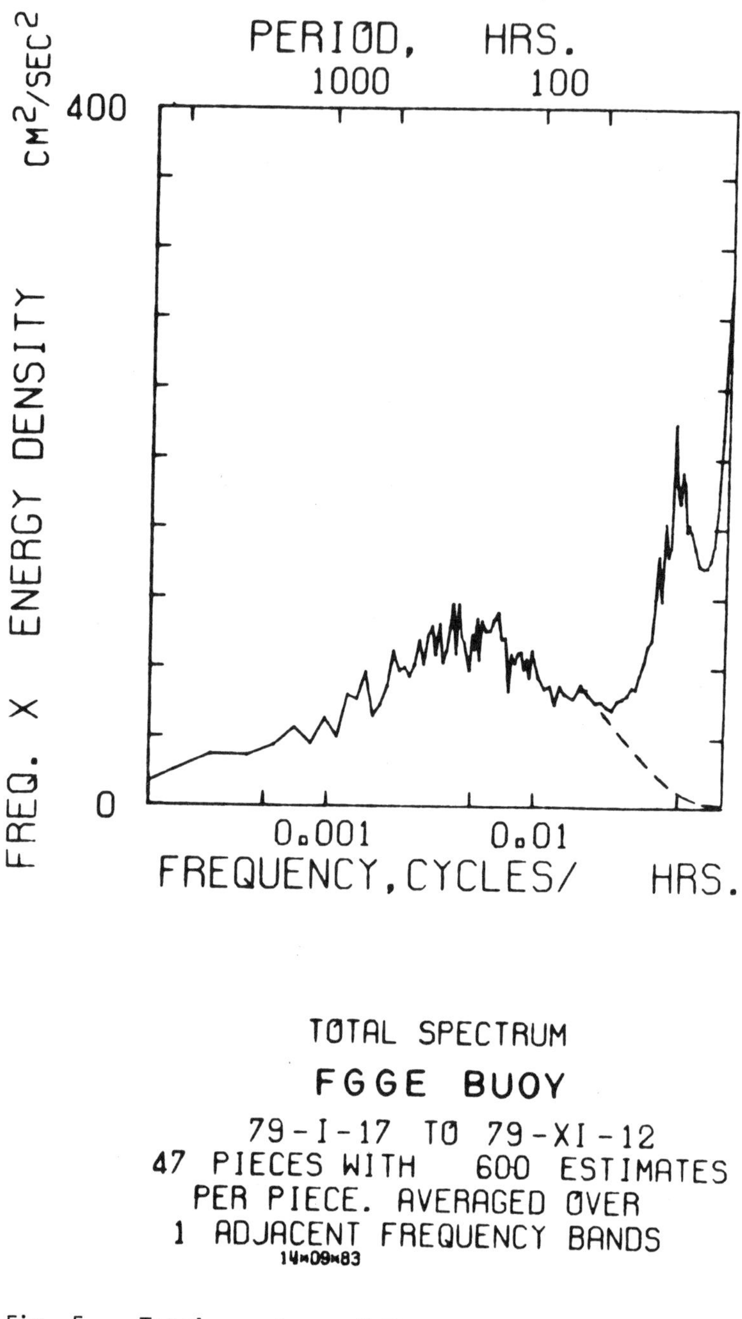

Fig. 5 Total spectrum of FGGE buoy trajectories

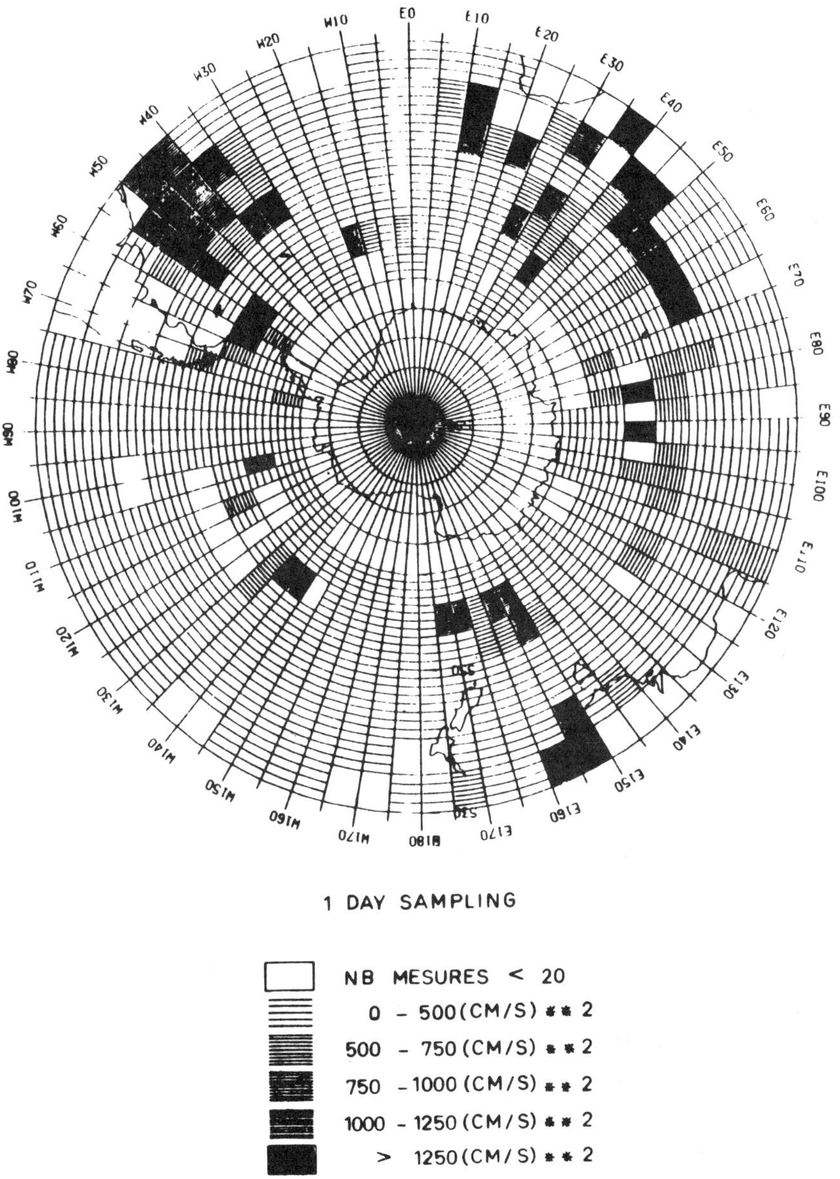

Fig. 6 Eddy kinetic energy from free drifting buoys, by $5^{o}x5^{o}$ cells

mean values in the Westerlies, and can only introduce some errors in the estimate of the surface mixed layer currents. Oceanic surface currents are the sum of several components which are geostrophic and ageostrophic in particular inertial, tidal, Ekman and Stokes ones.

From the European Center (Reading, UK) data, wind velocities have been interpolated at each buoy location and the corresponding wind spectrum is given in Fig. 7 by objective analysis. The coherence is larger in the anticlockwise domain (positive angular velocities) with an over 95% confidence level in the period range (2 days - 2 months); in the clockwise one, the coherence is smaller. From surface mooring measurements at site D (39° Northern Hemisphere) the coherence was found larger in the clockwise angular velocity domain (Gonella, 1972). The corresponding phase in the wind - buoy drift coherence is in agreement with the transient Ekman Theory: the mean phase (Fig. 8) gives a buoy drift at 17° on the left side of the direction of the 1000 mb wind. If we consider that in the Southern Hemisphere the surface wind is 15° on the left side from the 1000 mb wind direction, buoy drift directions are over 30° on the left side of the surface wind. This result is consistent with the one obtained by Mc Nally (1981) in the Northern Hemisphere; the current in the period domain ranging from 300 to 20 days is 3% of the corresponding wind velocities and decreases to 2% at 4 - 5 days period (Fig. 9).

Although this objective analysis is made in a Lagrangian frame, it shows that the surface mixed layer current in the period range (2 - 8 days of 50 - 200 h) is mainly ageostrophic. Recomputing the Eddy Kinetic Energy from buoy drifts series with a n days running mean (n varying from 2 to 9) the ratio E.K.E. (altimetry) over E.K.E. (drift) is converging to one (Fig. 10).

It is surprising that this 9 days value is similar to the barotropic time adjustment for the Southern Ocean; with a barotropic phase velocity of 2.000 km/day the Entire Antarctic Circumpolar Current is covered in about 10 days. This value is found by Wearn and Baker (1980) to be the time-lag between the mean wind stress all over the Antarctic Circumpolar Current and variations of the difference of the bottom pressure measurements across the Drake Passage.

Spectra from buoy drifts cannot be considered as Eulerian ones: the mean eastward velocity all over the A.C.C. estimated by buoy drift is of the order of 10 cm/s to 15 cm/s (10-15 km/day) and can affect spectra by Doppler bias mostly for wave length of 200 km and periods of several days. On the other hand, all the Seasat data analysis are not yet over; we are still computing Eulerian Spectra at the crossing points,

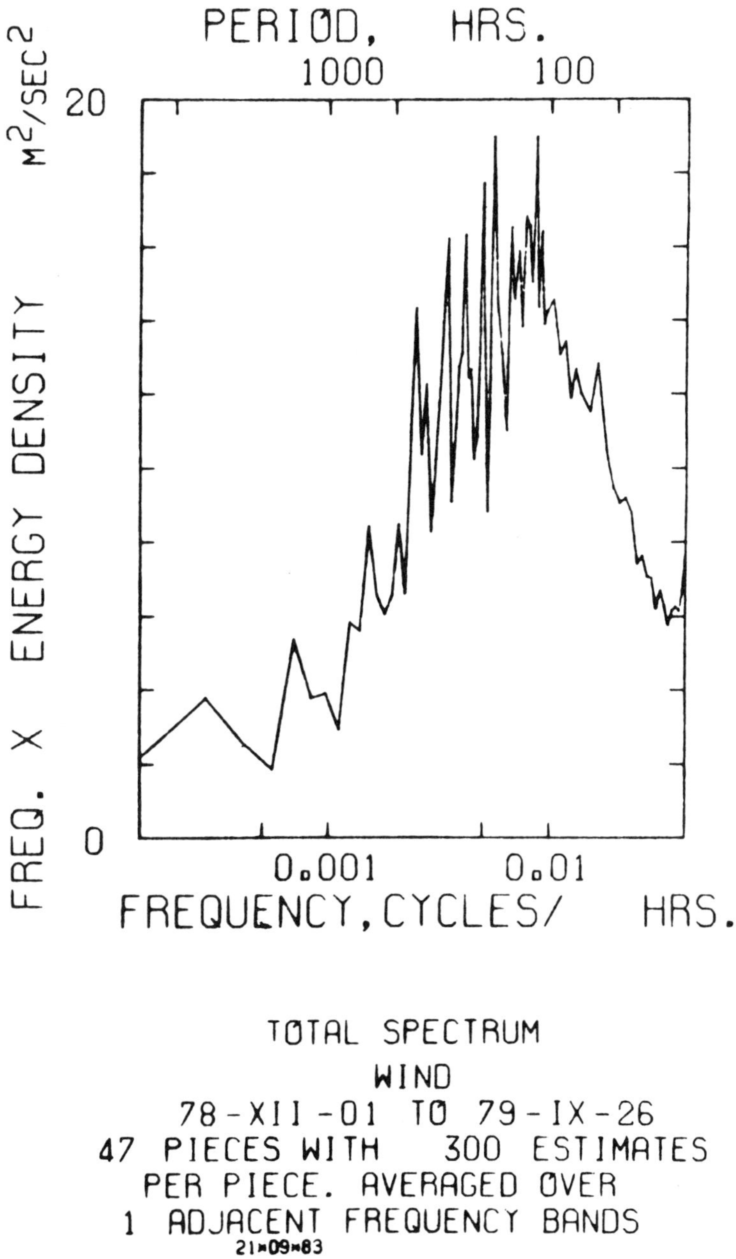

Fig. 7 Total spectrum of the wind at 1000 mb

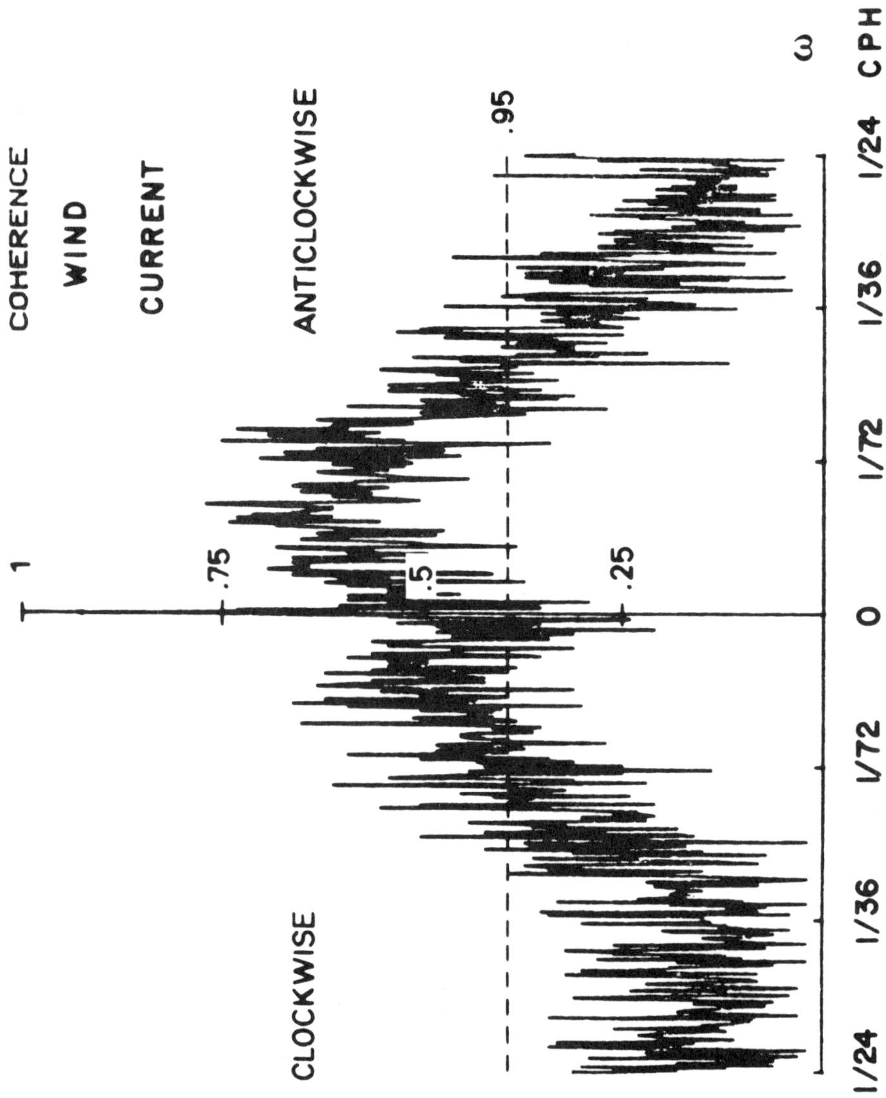

Fig. 8 Current to wind coherence

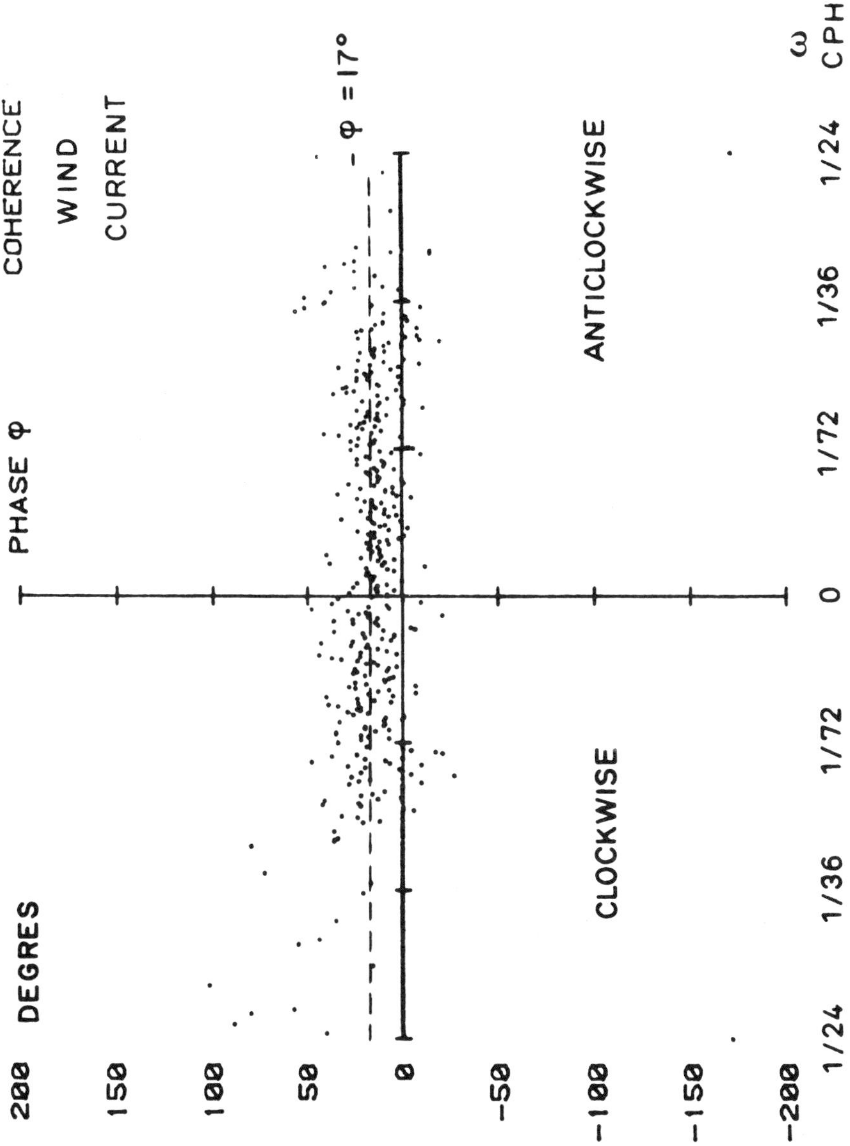

Fig. 9 Current to wind phasing

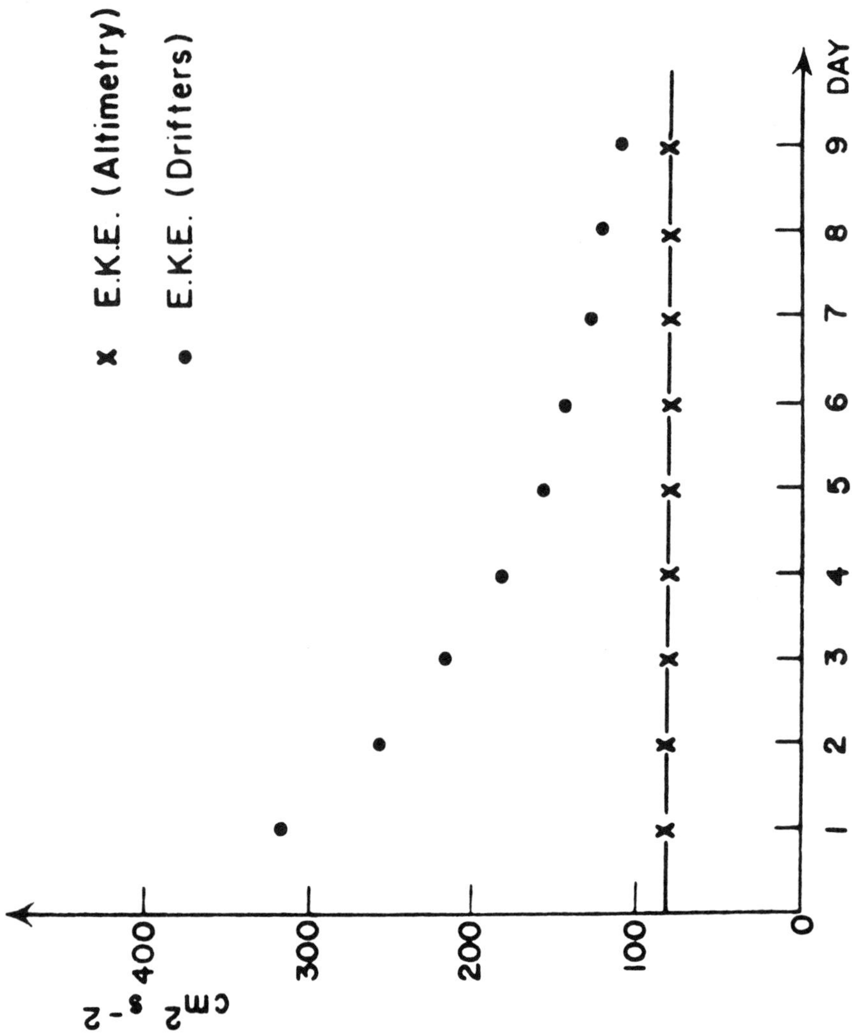

Fig. 10 Diagram of Eddy kinetic energy stability

estimating coherence in order to compare mean surface current with mean buoy drift.

5. CONCLUSION

Satellite technologies as Seasat and ARGOS system have demonstrated the capability to explore the dynamic of the global ocean.

The intercomparison which has been made between Seasat altimetry and FGGE buoy drifts shows the consistency of the two methods.

Nevertheless, drifters have to be considered as the best tracers we can have to estimate the surface current mainly in windy area and in Equatorial Zones. In these zones Ekman and inertial currents are strong and cannot be measured by altimetry. We have to point out (Gonella, 1983) that in equatorial area, inertial motions may have a mass transport not negligeable on several thousands of kilometers. Drifters represent a major tool for larger scale oceanographic experiments in addition to satellite altimetry and scatterometry.

REFERENCES

Cheney, R.E., J.G. Marsh and B.D. Beckley, 1983, Global Mesocale Variability from collinear tracks of Seasat Altimeter Data, J. Geophys. Res. 88, C7, 4343-4354.

Daniault, N., 1982, "Etude de la circulation oceanique superficielle dans la zone peri-Antarctique par l'analyse des derives de bouees de la P.E.M.G." Direction de la Meteorologie, Notes de travail de l'EERM 58, Avril 1983.

Fu, L.L., 1983, On the Wave Number spectrum of oceanic Mesoscale Variability observed by the SEASAT altimeter, J. Geophys. Res., 88, C7, 4331-4341.

Garret, J.F., 1980, Availability of th FGGE drifting buoy system data set, Deep Sea Research, 27 A, 145-159, 1980.

Gonella, J., 1972, A rotary component method for analysing meteorological and oceanographic vector time series, Deep Sea Res., 19, (12), Dec. 72, 833-846.

Gonella, J., 1983, "Mouvements d'inertie et structure zonale des courants oceaniques equatoriaux", Compte-rendu a l'Academie des Sciences de Paris (submitted Octobre 1983).

Lorenc, A.C., 1981, A global three dimensional multivariate statistical interpolation scheme. Monthly Weath. Rew., Vol. 109, 701-721.

Lutjeharms, J.R.E. and D.J. Baker, 1980, A statistical analysis of the mesoscale dynamics of the Southern Ocean, Deep Sea Res., 27 A, 145-159.

Mc Nally, G.J., 1981, Satellite tracked drift buoy observations of the nearsurface flow in the eastern mid-latitude North Pacific, J. Geophys. Res., 86, 8022-8030.

Menard, Y., 1983, Observations of Eddy fields in the Northwest Atlantic and Northwest Pacific by Seasat altimeter data, J. Geophys. Res., 88, C3, 1853-1866.

Richardson, Ph. L., 1983, Eddy Kinetic Energy in the North Atlantic from surface drifters, J. Geophys. Res., 88, C7, 4355-4367.

Wearn, R.B. and D.J. Baker, 1980, "Bottom pressure measurements across the Antarctic Circumpolar Current and their relation to the wind", Deep Sea Res., 27 A, 875-888.

SATELLITE MEASUREMENTS OF SEA-SURFACE TEMPERATURE FOR CLIMATE RESEARCH

P J Minnett, A M Zavody and D T Llewellyn-Jones

Rutherford Appleton Laboratory, Chilton, Didcot, Oxfordshire, UK

ABSTRACT

Factors determining the accuracy with which the sea surface temperature can be measured from space using infrared radiometers are discussed. These include the clear atmosphere radiative effects, contamination of the signal by clouds in the field of view, and sun glitter at the 3.7 μm 'atmospheric window'. The effects of near surface vertical temperature gradients caused by surface heat loss (the skin effect) and by solar heating (the diurnal thermocline) are also discussed. A review of present measurement capabilities is made and a brief description is given of the Along Track Scanning Radiometer (ATSR), which will fly on the European satellite ERS-1 to provide accurate measurements of sea surface temperature using a multi-angle, multichannel technique to overcome the effects of the clear atmopshere.

1 INTRODUCTION

The importance of precise global measurement of sea surface temperature (SST) to the aims of the World Climate Research Programme (WCRP-Houghton and Morel, 1983) has been stressed in many documents related to climate research (eg WCRP, 1981). It is not the intention here to further elaborate this theme, but rather to focus attention on some of the problems associated with achieving the required accuracies with known techniques (principally infrared radiometry) and foreseeable developments. We begin, however, by briefly stating the accuracy requirements of the WCRP and then summarise the measurement techniques and accuracies attained.

C. Gautier and M. Fieux (eds.), Large-Scale Oceanographic Experiments and Satellites, 57–85.

2 ACCURACY

The applications to which the SST measurements can be put depends critically on the accuracy with which they are made. The desired SST accuracy as a function of the spatial resolution of the measurement is shown in Figure 1, which is derived from a statement of requirement in WCRP, 1981. The absolute accuracy of the highest resolution measurements is relatively lax, the onus being on high relative accuracy to measure the structure of surface gradients. At larger scales the absolute accuracy requirement becomes more stringent.

Accuracies currently achieved with infrared measurements (see 4.1) are already useful for meso-scale studies, such as of the meandering and eddy formation at western boundary currents. The role of eddies in the oceanic general circulation and especially their heat transports is an important aspect of the WOCE (World Ocean Circulation Experiment) of the WCRP, and it is in this field that satellite SST measurements will be used with great effect in combination with radar altimeter data.

The highest accuracy, a few tenths of a degree Kelvin, is required for measurements on scales of several hundred to thousands of kilometres. This is needed, for example, for the TOGA programme (Tropical Oceans Global Atmosphere), which specified an accuracy of 0.3 K, to study the SST anomalies in the tropical oceans which have been shown by statistical methods (eg Barnett, 1981) to be related to the climate at mid-latitudes. The mechanisms of these 'teleconnections' is being revealed in studies of the atmospheric circulation using numerical models (eg Hoskins and Karoly 1981), which, as they increase in sophistication, will soon need realistic forcing from the air-sea interface. The first aim of the WCRP is to determine the predictability of climate and, if this is to be done for such 'teleconnections', accurate measurements of SST will be needed.

The second aim of the WCRP is to determine the extent of man's influence on climate, notably resulting from increasing the level of atmospheric CO_2. Madden and Ramanathan (1980) have discussed the difficulties in recognising a climate change in the long records of air temperature. This is due to the scarcity of suitable data, which is concentrated in the mid-latitudes of the Northern Hemisphere, and to the inherent variability of the climate system (weather). The global coverage provided by satellites ameliorates the sampling problem, and, provided they can be made accurately (specifically, without biases), SST measurements from satellites could provide a useful monitor of climate change. Since the changes are slow in comparison with the life-time of a satellite instrument, measurements from many systems over decades are

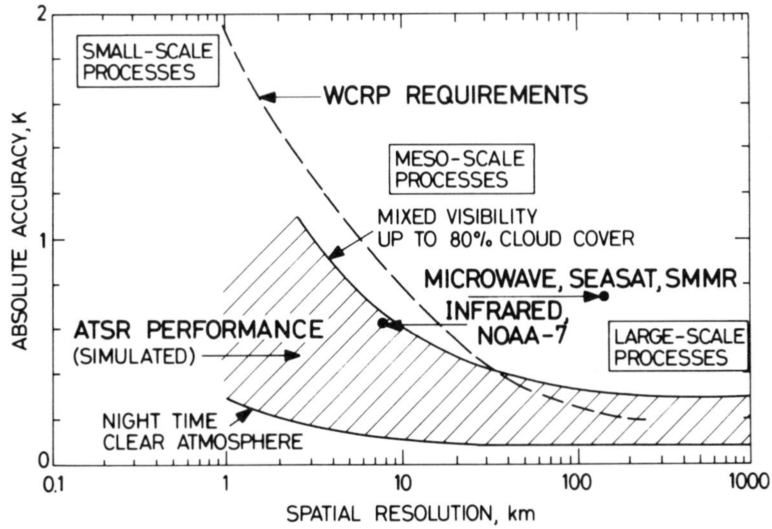

FIGURE 1
Accuracy requirements for SST measurements in the WCRP (dashed line). Currently attainable accuracies are shown for the NOAA-7 AVHRR/2 and the Seasat SMMR (see 4), and the simulated accuracy of ATSR (see 6).

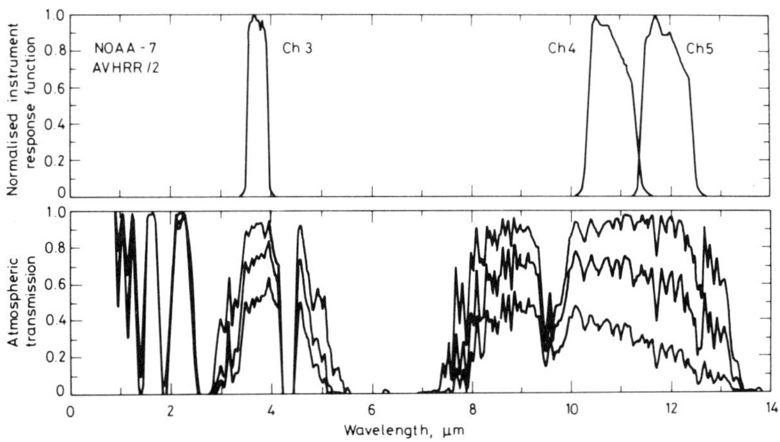

FIGURE 2
Theoretical spectra of atmospheric transmission, at nadir, in the infrared region at 1 to 15 μm wavelength. The three spectra correspond to different amounts of precipitable water (7 mm - polar, 29 mm temperature, 54 mm tropical). The response functions of channel 3 (3.7 μm), channel 4 (11 μm) and channel 5 (12 μm) of the AVHRR/2 on the NOAA-7 satellite are shown, and indicate the 'atmospheric windows' used. (After Llewellyn-Jones et al, 1983).

needed, thus stressing the importance of measurements of high absolute accuracy.

3 MEASUREMENT TECHNIQUES

Sea surface temperature has been measured from satellites using infrared and microwave radiometers. Each offer distinct advantages but also suffer distinct disadvantages. These will be discussed briefly for each in turn.

3.1 Infrared Measurements

The atmosphere is translucent to infrared radiation in a few wavebands, referred to as 'atmospheric windows'. These are shown in Figure 2 in spectra calculated for three atmospheres with different concentrations of water vapour content together with the instrument response functions of the Advanced Very High Resolution Radiometer Mk2 (AVHRR/2) of the NOAA-7 satellite, which operates in the 'windows' at 3.7, 11 and 12μm.

The transmittance of the 'windows' is dependent not only on the water vapour concentration but also on its vertical distribution. Since this dependence varies with wavelenth a method of determining the transmittance of the atmosphere is facilitated by making co-registered measurements in different 'windows' (Prabhakara et al, 1974; McMillin, 1975; Deschamps and Phulpin, 1980). This is the basis of infrared multichannel SST measurement.

The measurements in the infrared windows are advantageous for SST determination because they are close to the maximum in the thermal emission spectrum at terrestrial temperatures (especially the 10 to 13 μm window) and that the emitted energy is strongly dependent on the temperature of the emitter ($\sim T^{13}$ at $\lambda = 3.7$ μm, $\sim T^5$ at $\lambda = 11$μm). Futhermore the emissivity of the ocean surface is very close to unity and insensitive to surface conditions.

From the instrumental aspect, absolute calibration targets (black bodies) are relatively easily achieved and current optical systems can provide good spatial resolution (~ 1 km).

However, in addition to the effects of the clear atmosphere, infrared techniques suffer the very big disadvantage that clouds are not transparent. Thus a significant part of most images of the sea surface is obscured by cloud. The measurements are also easily contaminated by atmospheric aerosols. Methods of identifying clouds are discussed in 5.3.

Given the temperature(s) of the on-board calibration target(s) and knowledge of the filter profiles, the telemetered signal can be converted to brightness temperatures (T_i) for each channel (eg Lauritson et al, 1979). The SST is obtained from a linear combination of the brightness temperatures :

$$SST = a_0 + \sum_{i=1}^{N} a_i T_{\lambda i}$$

Where N is the number of channels used. The coefficients a_0, a_i are determined, empirically or theoretically, to give the optimum performance in a given set of atmospheric conditions believed to represent fairly those in a particular area, or period, or for the whole globe. The correct determination of the coefficients is critical for the accurate measurement of SST from satellites (see 5.1).

3.2 Microwave Measurements

The great advantage offered by microwave measurements is that most clouds are transparent, with problems arising only in clouds with large water droplet size and concentrations. Thus an almost all-weather global measurement of SST is, in principle, possible with microwave techniques. The disadvantages suffered by microwave sensors, when compared with those in the infrared, include low levels of emitted energy at terrestrial temperatures which in turn shows only a linear temperature dependence, a sea surface emissivity which is less than unity and dependent on many factors such as surface conditions, temperature, frequency, polarisation and viewing angle. With present day antenna sizes the spatial resolution is comparatively coarse (150 km for the Scanning Multi-channel Microwave Radiometer - SMMR - on Seasat and Nimbus-7) with complicating patterns of side-lobes. The poor resolution need not be, in itself, a hindrance for climate studies, but the presence of land in the footprint or side-lobes can severely contaminate the measurement. The complicated emissivity dependence can be turned to advantage in that microwave radiometry can be used to measure other variables, such as wind speed (eg Njoku, 1982).

The conversion from telemetered signal to brightness temperature is somewhat more complicated than in the case of the infrared measurements, but the final SST retrieval consists of a combination of brightness temperatures from differing channels and polarisations. The algorithms can be complicated and are generally derived from theoretical considerations (eg Wilheit and Chang, 1980), although empirical adjustments are sometimes made. Some of the shortcomings of existing satellite microwave SST measurements are due to instrumental effect which may be avoided in future designs of microwave radiometers. Indeed it is arguable that, on account of the instrumental difficulties inherent in the design and operation of SMMR, it has not yet been possible to carry out a comprehensive experimental appraisal of microwave radiometry for SST measurement.

4 ACCURACIES ATTAINED

Because satellites provide a new type of measurement, an instantaneous average over a given area, there is no reliable 'conventional' data with which they can be compared to estimate their accuracy. In the absence of any real alternative nearly all estimates of accuracy have been made using in situ data, generally sub-surface point measurements taken from a ship or a buoy. Inherent in this method are sources of error due to vertical gradients in the water column (see 5.5, 5.6), horizontal variability within the area of the satellite measurement and, since few in situ measurements have been coincident with satellite overpasses, also temporal changes. Consequently, with these sources of error in addition to the uncertainty in the in situ measurement itself, the figures thus derived are pessimistic in describing the accuracy with which the satellite makes its measurement.

TABLE 1

PUBLISHED ESTIMATES OF ACCURACY OF INFRARED SST MEASUREMENT (all for AVHRR)

Author	Region	Channels (μm)	Day or Night	Mean Error (K)	Scatter (K)
Bernstein (1982b)	Pacific	3.7, 11	Day	- 0.10	0.54
McClain et al* (1982)	Global	3.7, 11	Night	0.20	0.62
		11, 12	Day	0.02	0.78
		11, 12	Night	- 0.03	0.63
		3.7, 11, 12	Night	0.02	0.58
Llewellyn-Jones et al (1983)	North Atlantic	11, 12	Day	- 0.10	0.53

* These authors express scatter as RMSD, others as RMS. The values are equivalent for zero mean error. The higher values of scatter reflect the wider range of atmosheric conditions considered; the low mean errors result from the use of an empirical bias correction.

4.1 Infrared Accuracies

The mean error and scatter of the differences between satellite and in situ SST measurements are shown in Table 1. The algorithm tested by Bernstein was derived empirically by a regression analysis between brightness temperatures and in situ measurements. The figures shown result from testing this algorithm with an independent set of in situ data. The algorithms of McClain et al have been empirically adjusted and that of Llewellyn-Jones et al was derived entirely from modelling the radiative transfer through the atmosphere. The values are all just more than half a degree. If one can assume that the differences between the satellite and in situ temperatures arise equally from uncorrelated errors in the satellite measurement, the in situ measurement and the imperfect method of comparison, then the error attributable to the satellite measurements alone can be given as $1/\sqrt{3}$ of the values in the table. Thus a 0.6 K total error has a component due to the satellite measurement of 0.35 K. This is close to the requirements of the WCRP.

4.2 Microwave Accuracies

Taylor (1983) has summarized the results of comparisons between the Seasat SMMR SST retrievals, using different algorithms, and different sets of in situ data. He finds that the mean bias in ten studies is -0.1 K, with a mean value of the standard deviation of the errors of 1.0 K.

In an attempt to reduce the errors arising from comparing point measurements with satellite observations over 150 km footprints, Bernstein (1982a) compared time averaged gridded ship data from the Pacific Ocean with SMMR data on the same grid. He found a mean error of 0.22 K and standard deviation of 0.8 K.

While microwave observations offer cloud penetration, their accuracy at present is not satisfactory for WCRP purposes.

5 PROBLEM AREAS

The oceanographic elements of the WCRP in the period 1988 to 1993 will depend heavily on satellite infrared radiometers for the measurement of SST. The AVHRR will be included on the TIROS-N weather satellite series of NOAA scheduled for this period, and these will be augmented by the Along Track Scanning Radiometer (ATSR, see 6), an experimental instrument on the European ERS-1 satellite. The only microwave instrument, the SMMR on Nimbus-7 will have been in orbit for nearly a decade in 1988 and so its reliable performance through the period is quite unlikely. Consequently it is worth concentrating on some aspects which need serious attention if satellite infrared measurements of SST are to be fully relevant to the WCRP.

These factors can be divided into two groups : firstly, those that contaminate the infrared measurement of the surface 'skin' temperature, and secondly, those which cause vertical temperature gradients within the ocean surface layer which will lead to errors if the radiometric skin measurement is interpreted as a bulk 'SST' measurement comparable to those taken by conventional in situ methods.

The first group includes the emission, absorption and scattering effects of the clear atmosphere in the field of view, the effects of clouds and aerosols, the effects of reflection of solar radiation by the sea surface into the field of view and the effect of the variability of the emissivity of the sea surface. The second group includes the 'skin effect' and the diurnal thermocline.

Since the amount of data that scanning radiometers produce is very large (the data rate of AVHRR/2, with five channels at full resolution, is 0.66 Mbps) and consequently all algorithms used to process these data in an efficient manner must be simple as well as correct.

5.1 Clear Atmosphere Effects

The algorithms for retrieving SST using multichannel measurements, as described in section 4 and elsewhere in these proceedings, can be based on simulations using a radiation transfer model (eg Llewellyn-Jones et al, 1983), by an empirical analysis of coincident satellite and in situ measurements (eg Bernstein, 1982b) or by a combination of the two. In all cases the success of the algorithm, where applied to measurements made in a wide variety of atmospheric conditions, depends on how well these conditions have been represented in the algorithm derivation. For satellite measurements made in the atmospheric 'windows', the main radiatively important component exhibiting significant variability is water vapour, and in the radiative transfer modelling this variability is represented by a set of profiles of atmospheric temperature and water vapour concentration derived from radiosonde ascents. In the empirical approach, the coincident satellite and in situ measurements must be made in a sufficient variety of conditions. In both methods it is difficult to ensure that the range of conditions is adequately represented, given the scarcity of suitable observations, especially over the oceans of the southern hemisphere.

Satellites, of course, do provide a method of obtaining global measurements and the satellite soundings of temperature and humidity should be incorporated in schemes to derive globally applicable SST retrieval schemes.

One possible way of improving current operational SST retrievals from AVHRR is to incorporate a factor describing the change in path length across the scan. The atmospheric path length, or airmass, varies as secant(ψ) for $\psi < 60°$ where ψ is the satellite zenith angle. Llewellyn-Jones et al (1983) show that the inclusion of an airmass dependency, either explicitly or through a secant(ψ) approximation (as proposed by McClain and Walton, 1982) can significantly reduce the scatter and bias error in SST retrievals when compared with oceanographic research ship measurements. Furthermore, it extends the proportion of the AVHRR scan which can be used for quantitative SST measurement from 1440 km, using current algorithms, to 2280 km.

5.2 Sun Glitter at 3.7μm

The sun glitter pattern, caused by the tilting of facets of the sea surface by waves thus spreading the point of specular reflection of the sun over a large area, was related to the local wind speed by Cox and Munk (1954). Their observations were in the visible part of the spectrum, but the same effect occurs in the near infrared where the emitted radiation at sea temperatures is small (0.56 W m^{-2} st^{-1} μm^{-1}, λ = 3.7 μm, T = 15 °C). In the thermal infrared the emitted power (7.99 Wm^{-2} st^{-1} μm^{-1}, λ = 11 μm, T = 15 °C) is much greater than the reflected solar radiation, so that the sun glitter contribution is negligible.

The glitter intensity at 3.7μm across a mid-latitude scan of AVHRR/2 on NOAA-7 is shown in Figure 3 for a range of wind speeds. The sun-earth-satellite geometry corresponds to the southern edge of the image in Figure 6. The effect of the glitter is to introduce positive bias in values of SST derived using measurements from AVHRR channel 3. If, for climate research, bias errors from this source are to be kept below 0.1 K (in addition to all other uncertainties) then the reflected solar power must, in general be <1 mW m^{-2} st^{-1} μm^{-1}. The proportion of the scan of AVHRR/2 in a NOAA-7 orbit (local time ~ 1430 hours) in which this level is exceeded, is shown in Figure 4. The area over which 3.7 μm data cannot be used for accurate determination of SST, without employing an accurate sun glitter model, can be large for even moderate wind speeds.

5.3 Clouds

Although accurate descriptions of cloud cover and cloud types

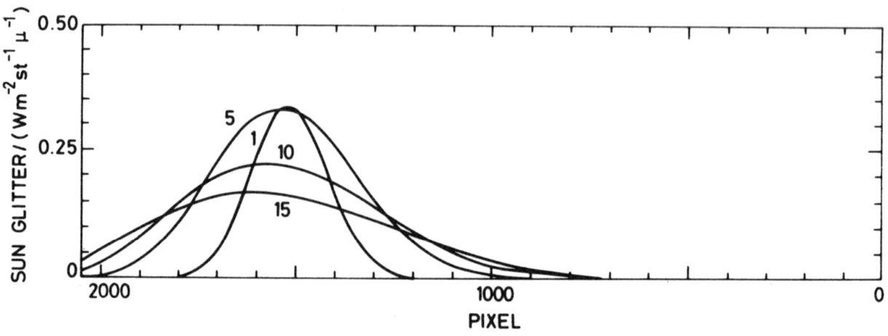

FIGURE 3
The intensity of the sun glitter pattern at λ = 3.7 µm, for a range of surface wind speeds, calculated using the Cox and Munk (1954) relationship. This case corresponds to the southern edge of Figure 6, which is from pixel 1024 (nadir) to 1536.

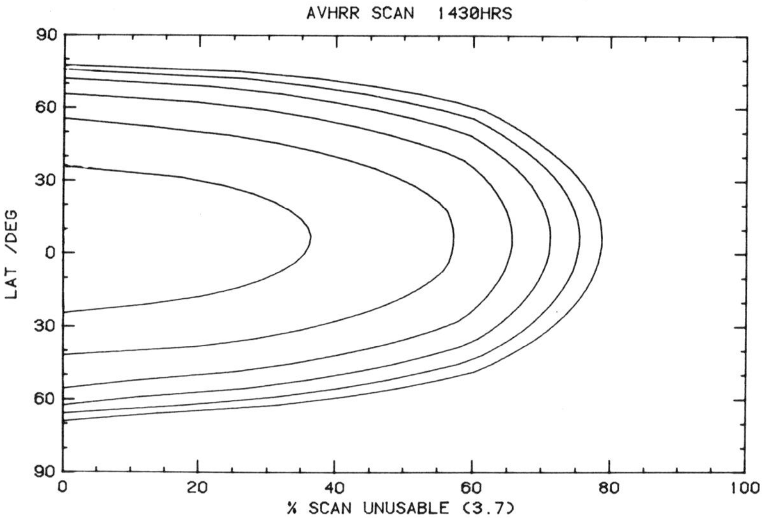

FIGURE 4
The proporion of a NOAA-7 AVHRR/2 scan for which the sun glitter radiance at λ = 3.7 µm is more than 1 mW m^{-2} st^{-}m^{1} µm^{-1}, as a function of latitude. The wind speed dependence is derived from the Cox and Munk (1954) relationship and it is assumed that each scan is over cloud free ocean, without land, and that the wind speed is spatially uniform. The calculations were made for the case of the equinoxes. The effect of the changing solar declination angle through the seasons is to shift the pattern with respect to latitude.

are themselves meteorological variables of prime importance to the WCRP they will be considered here only as a contaminant in measurements of sea-surface brightness temperature. The removal of their effects is necessary for accurate SST determination.

The methods so far employed to identify cloud contaminated pixels can be grouped in increasing complexity, starting with the simplest which involves, for example, comparing the differences between co-registered pixels from different channels with a threshold value. The next group includes techniques involving the local statistics of brightness temperatures as measured in individual channels. More complicated methods included in the final group are those involving the inversion of sets of non-linear equations whereby not only is the SST retrieved, but also cloud amount and temperature (Dozier, 1981). The recent method of Arking and Childs (1983) involves using models of the radiative properties of cloud types and is aimed at deriving cloud parameters from the multispectral measurements of AVHRR. Once these parameters are obtained, however, one could proceed to retrieve the SST itself.

Remembering the need for algorithms to be fast, and hence simple, only the techniques of the first two groups will be discussed here.

5.3.1 Threshold Methods. The majority of cloud types have a high albedo in the visible and very near infrared, in contrast to the sea surface which, except in areas of sun glitter, appears very dark. Consequently channels 1 (0.57 μm to 0.69 μm) and 2 (0.71 μm to 0.99 μm) of AVHRR are very sensitive cloud detectors. Even pixels with small amounts ($<10\%$) of thick cloud can be readily identified. This method can only be used in the daytime half of each orbit and is complicated by the changing threshold level across the scan due to changing sun-earth-satellite geometry (see Figure 7a).

In the atmospheric 'windows' the differences in the brightness temperatures of the sea-surface are limited to a range of values less than 4K. Pixels in which these values are exceeded must be contaminated by cloud or land (unless there is a co-registration error between the channels). This method can be used for data from both night and day but again is complicated by changing values of the thresholds across the scan (see Table 2). While values outside the thresholds are a sufficient condition for the identification of contamination, the reverse is not necessarily true; ie a value lying between the thresholds is not a guarantee that the pixel is free of contamination (see Figure 1 of McClain et al, 1983).

5.3.2 Statistical Methods. The two methods discussed here are the Truncated Normal Distribution technique (Crosby and Glasser, 1976) and the Spatial Coherence technique (Coakley and Bretherton, 1982). Both can be applied equally well to daytime and night-time data.

The first method assumes that the SST is uniform in a chosen area of partially cloud filled image, and is warmer than the clouds. The warm end of a histogram of the pixel temperatures is then a truncated Gaussian, with a half-width defined by the radiometer noise level. Cooler pixels are deemed to be contaminated by clouds. If the radiometer noise level is known, a Gaussian curve can be fitted to the warm end of the distribution, with the central value giving the required SST. The method requires there to be small SST variability over quite large areas, and runs into difficulties when land is included in the scene.

The second method assumes that SST fluctuations in a small area are much less than the changes in temperatures recorded at cloud edges, or than fluctuations in cloud top temperature. The variances in small sets of pixels (say from a 3 x 3 array) are calculated and arrays are rejected when these exceed a preset threshold value (say 0.01 K^2). This value can be constant across the scan and for most types of scenes. The method may fail with stratiform clouds with near uniform temperature and with very low cloud, or fog, showing small temperature contrast to the underlying sea.

The performance of these methods is compared by considering two case studies :

5.3.3 Case Study A : Comparison with in situ measurements. Some fifty cases of in situ measurements in the North Atlantic, mostly from oceanographic research vessels, made within two hours of a NOAA-7 overpass have been analysed by Llewellyn-Jones et al (1983) as a test of an SST retrieval algorithm derived entirely from radiative transfer modelling. The data are from daytime measurements in July to September 1981 in a variety of cloud conditions. The SST values were retrieved using 'split-window' measurements (Channels 4 and 5 on AVHRR/2; thus avoiding the need to consider sun glitter distribution), from a 50 x 50 pixel array centred on the in situ measurement. Pixels contaminated by cloud were identified using a selection of the methods described above. In particular a comparison can be made between the Truncated Normal Distribution method (TND) and the Spatial Coherence method followed by a Visible Threshold discriminator (SCV).

TABLE 2

SHOWING THE MAXIMUM TEMPERATURE DIFFERENCES BETWEEN THE INFRARED CHANNELS ON THE NOAA-7 AVHRR/2 DUE TO CLEAR ATMOSPHERE EFFECTS

Airmass	Zenith Angle	$T_{3.7} - T_{11}$	$T_{11} - T_{12}$
1.00	0°	-0.6 to 3.7	0.0 to 3.3
1.25	36.9°	-0.7 to 3.6	0.0 to 3.5
1.50	48.2°	-0.7 to 3.8	0.1 to 3.6
1.75	55.2°	-0.7 to 3.8	0.1 to 3.8
2.00	60.0°	-0.8 to 3.8	0.2 to 3.8

Where T_x is brightness temperature at $\lambda = x$ μm

These values are from a radiation transfer model using the NOAA-7 AVHRR/2 infrared instrument transfer functions (see Figure 1) and a set of radiosonde profiles describing global atmospheric variability. The model is described in Zavody, 1982, and briefly in Llewellyn-Jones et al, 1983.

TABLE 3

COMPARISON OF ERRORS, RESULTING WITH DIFFERENT CLOUD CLEARANCE SCHEMES, DETERMINED AGAINST IN SITU MEASUREMENTS

Cloud Clearance	Mean Error	Sd of Error
SCV	-0.10 K	0.53 K
TND	0.24 K	0.59 K

(after Llewellyn-Jones et al, 1983)

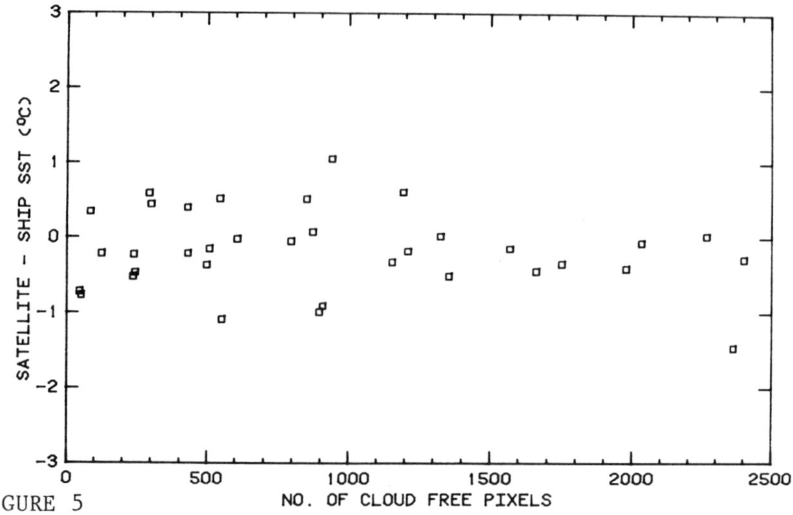

FIGURE 5
The difference between satellite SST measurements and corresponding in situ measurements from oceanographic research vessels, shown as a function of cloud cover. The satellite values are from 'split-window' measurements (11, 12 μm) derived for areas of 50 x 50 pixels after cloud clearance using Spatial Coherence plus Visible Threshold methods. (From Llewellyn-Jones et al, 1983).

Of the 51 cases analysed, the TND method failed to retrieve and SST value in 12 of them, due to large SST gradients and heavy cloud cover. The SCV method used a threshold variance value of 0.01 K^2 over 3 x 3 pixel arrays followed by a visible albedo threshold determined from the local statistics of the visible channel measurements. The mean and standard deviation of the errors for the 39 cases where a comparison was possible are shown in Table 3. In particular, the mean error of the TND method is worse.

The differences between the in situ measurement and the SCV SST are shown in Figure 5 as a function of cloud cover. No significant trend is seen.

5.3.4 <u>Case Study B</u> : NOAA-7 orbit 903 on 26 August 1981. Figure 6 shows an area (512 x 512 pixels) to the North West of Madeira (island bottom right). The scene is not corrected for geometrical distortion and the sub-satellite track is along the right hand edge. This example was chosen because of the range of cloud types and cover, the presence of a strong sun glitter signal and because the SST variability in this area might be small (ie it is away from major current boundaries, Gulf Stream Rings, and coastal effects).

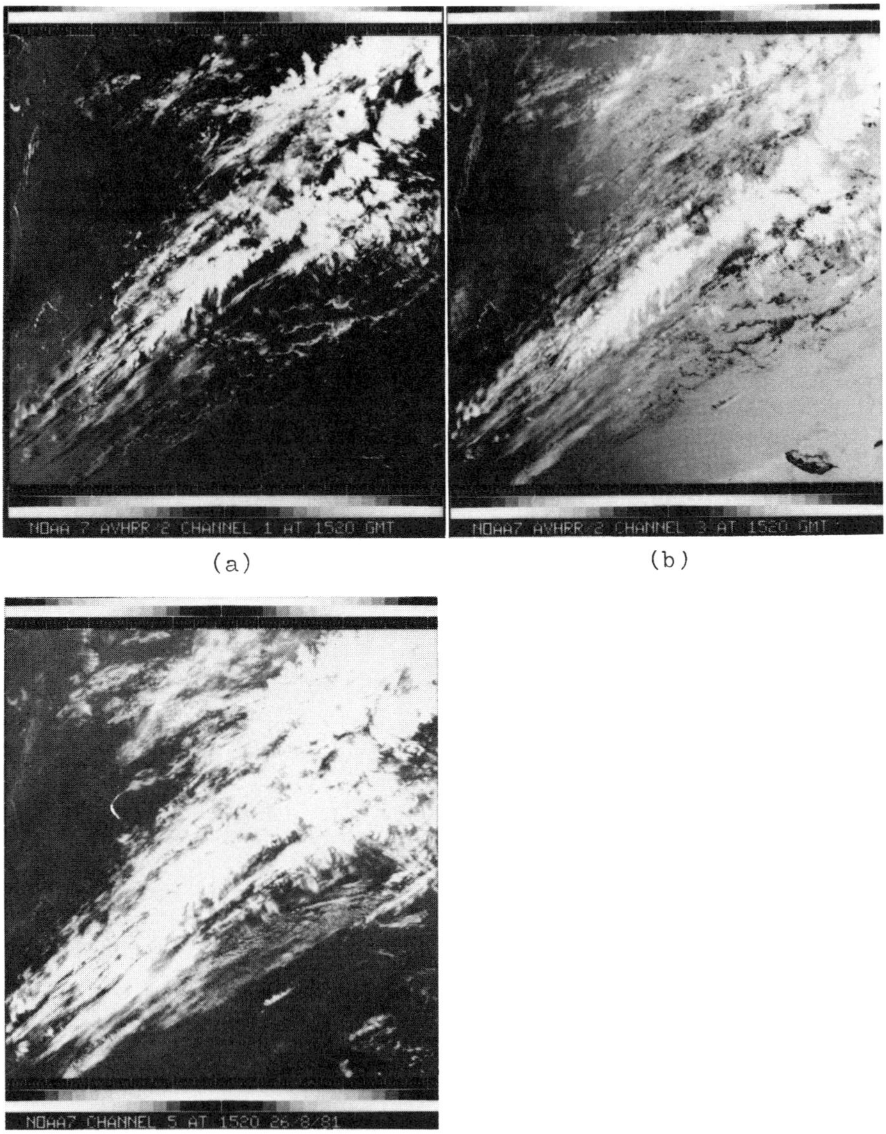

(a) (b)

(c)

FIGURE 6
Data from the AVHRR/2 on NOAA-7 on 26 August 1981 (orbit 905), shown in image form for (a) Channel 1, (b) Channel 3, (c) Channel 5. The area shown in 512 x 512 pixels (not corrected for geometrical distortion) to the North West of Madeira, which is the island at lower right. The sub-satellite track is at the right hand edges of the figures, from which true North is about 11° in clockwise direction.

The low level of cloud cover at the southern edge of this scene means that this part can be used to investigate the effects of the pronounced sun glitter pattern. Figure 7a shows histograms of digital counts from the visible channel for 50 x 50 pixel arrays along the southern edge of the scene. The shift is due to increasing sun glitter. The values of the visible threshold can be derived by applying a TND algorithm to the dark end of a histogram of visible counts. The threshold values for cloud contamination derived this way agree with those obtained independently by manual inspection. The SST values derived from a 'split-window' (11 and 12 μm) algorithm (Llewellyn-Jones et al, 1983), without cloud clearance, are given in Figure 7b, showing that 'split-window' SST retrievals are independent of sun glitter contamination.

Dividing the whole scene into 100 50 x 50 pixel arrays provides a set of examples with cloud cover ranging from almost zero to almost 100%. Cloud cover is defined here as the proportion of pixels with visible digital counts greater than the locally determined threshold value. For arrays with almost total cloud cover, threshold values interpolated from neighbouring arrays were used. This estimation of cloud cover is not accurate for purposes of cloud cover study itself as no allowance is made for pixels partially filled with cloud. It is, however, adequate for present purposes.

By using different cloud clearing methods on these pixel arrays, an indication of their relative performances in different conditions of cloud cover can be obtained. The 'split-window' SST values were calculated for each array after cloud clearance using :

a/ the Visible Threshold method with local threshold values,
b/ the Spatial Coherence method with variance thresholds ranging from 1 to 0.01 K^2,
c/ the TND method.

FIGURE 7
Histograms from 50 x 50 pixel arrays of (a) the visible signal (Channel 1) and (b) the computed 'split-window' (11, 12 μm) SST, (without cloud clearance). The arrays are taken from the southern edge of the area shown in Figure 6. The effect of increasing sun glitter intensity is shown clearly in the visible channel data, yet the 'split-window' SST remains unaffected. The consequence of warm land is to be seen in the temperature histograms (b).

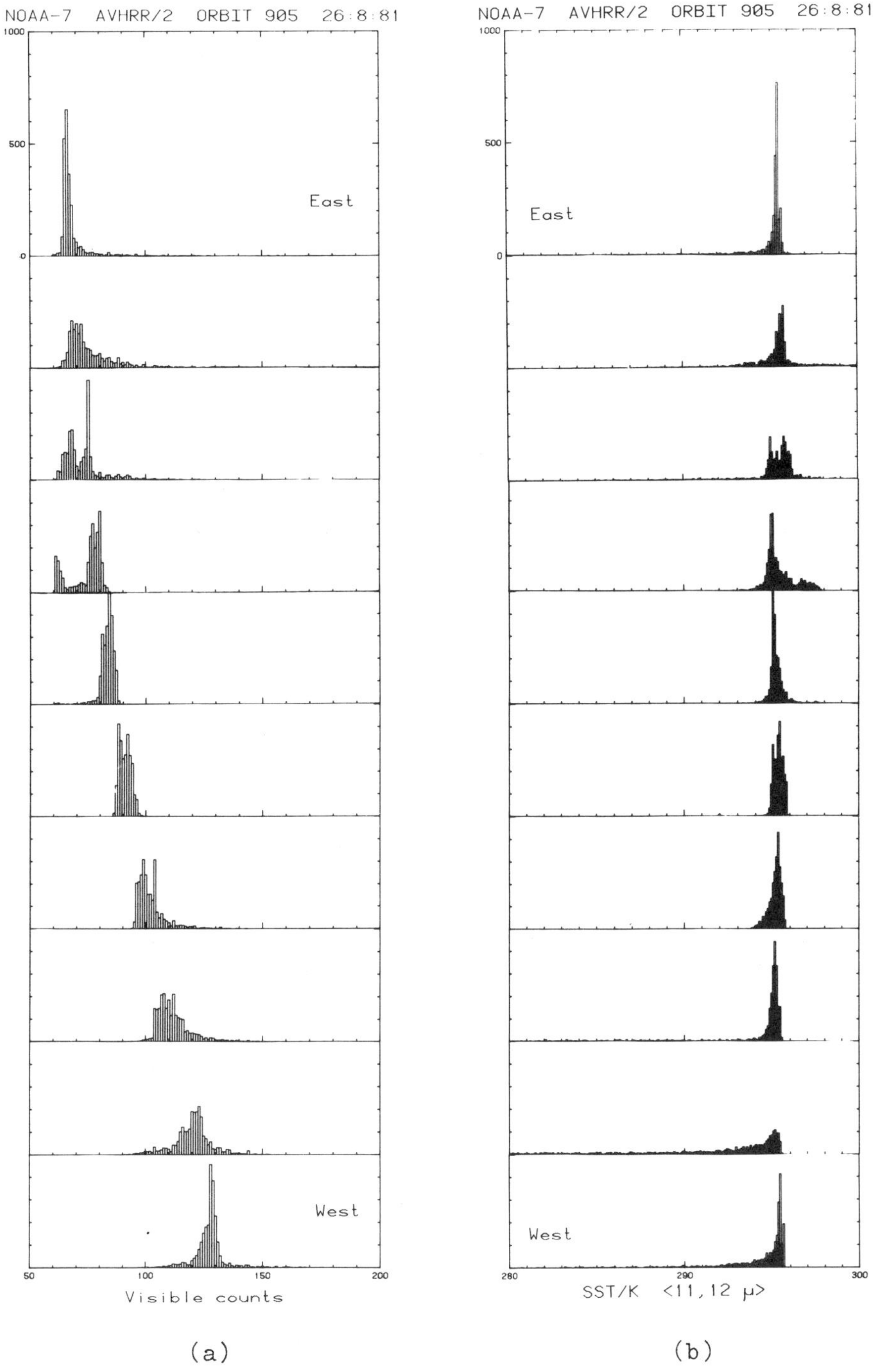
NOAA-7 AVHRR/2 ORBIT 905 26:8:81
1000
500
0
East
West
50
100
150
200
Visible counts
NOAA-7 AVHRR/2 ORBIT 905 26:8:81
1000
500
0
East
West
280
290
300
SST/K <11, 12 μ>

(a)

(b)

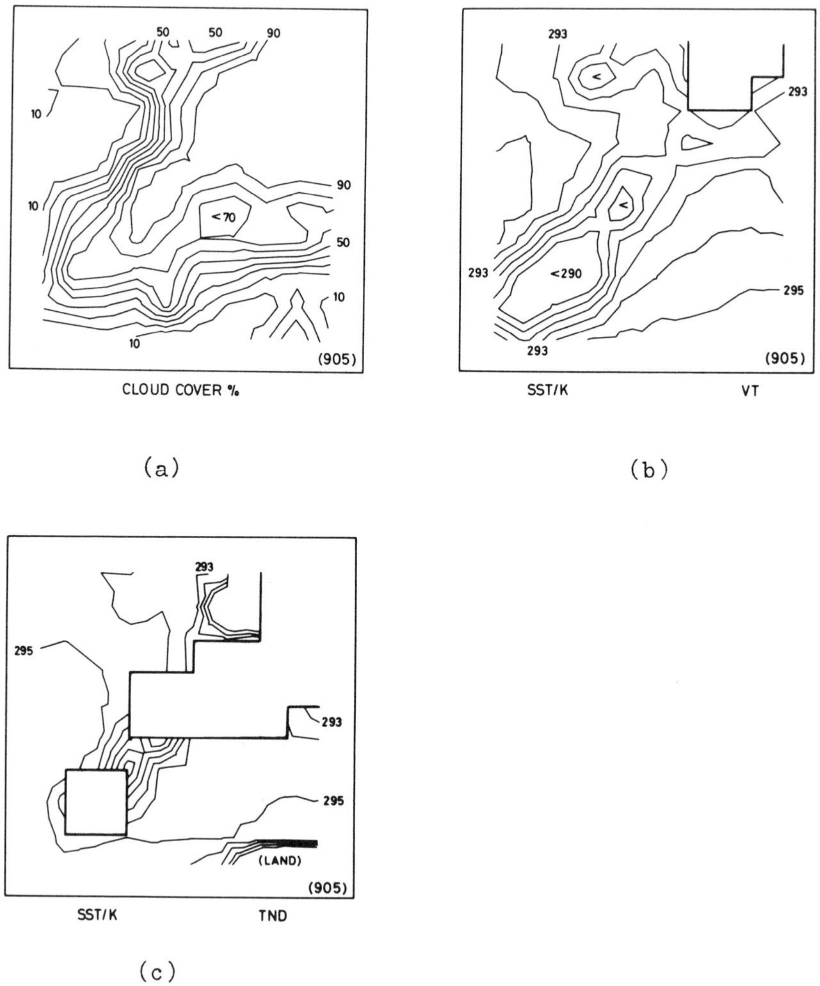

FIGURE 8
Maps produced from the data shown in Figure 6 (orbit 905) depicting (a) cloud cover; (b) the 'split-window' SST field (11,12 μm) derived after cloud clearance using a visible threshold technique; (c) as (b) but with cloud clearance using the Truncated Normal Distribution method. Values were derived from 50 x 50 pixel arrays. Contour intervals are 10% in(a), 1K in (b) and (c). The shaded areas in (b) and (c) are where cloud cover prevented a determination of SST. Obvious similarities exist between the SST fields and the cloud cover.

(a)

(b)

(c)

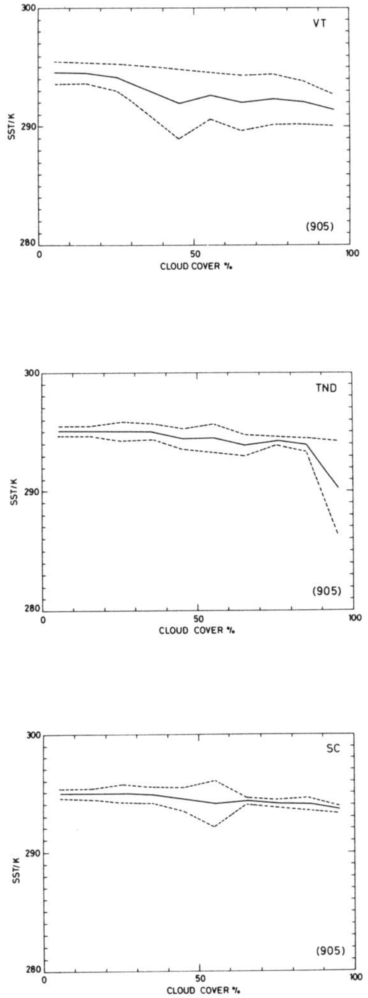

FIGURE 9
Showing the dependence of derived SST on cloud cover. The 'split-window' SST (11, 12 µm) were derived for 50 x 50 pixel arrays from the data in Figures 6 and 8, with cloud clearance by a) the Visible Threshold method; b) the Truncated Normal Distribution method; c) the Spatial Coherence method with a variance threshold of 0.01 K^2. The SST values were grouped according to cloud cover (10% increments) and mean SST (± 1 standard deviation) are plotted. All show a trend of decreasing SST values with increasing cloud cover.

The resulting SST fields were automatically contoured, along with the values of cloud cover. All SST fields show similarity with the cloud cover field in that a tongue of cold SST values lies along the axis of the high cloud cover. The coldest SST retrievals coincide ith a region where the clouds are particularly fragmented (see Figure 8 where the TND and Visible Threshold fields are shown as examples). Plotting the retrieved SST values as functions of cloud cover shows obvious trends in all cases (Figure 9). The sizes of the change in SST with cloud cover are too large to be explained by cloud cover modulating the diurnal heating, especially given that the sun glitter pattern is consistent with a surface wind speed of order 15 ms^{-1}. Even though the sample is small (there are 100 independent 50 x 50 pixel arays in this area, but this number was reduced by some cases in which the cloud cover was so thick that the cloud clearance schemes failed to permit the retrieval of an SST value), the conclusion must be that none of these simple cloud clearance algorithms is performing adequately to give SST values to the accuracy required by the WCRP in cases of 'cloud cover' above say 40%. The Spatial Coherence method with a threshold value of 0.01 K^2 seems to be the technique most insensitive to cloud cover. Admittedly these examples can be much improved by 'tuning' the algorithms for optimal performance in each array, but this is not desirable for the routine processing of large amounts of data for WCRP purposes. Consequently, in order to meet the WCRP requirements more reliable cloud clearance schemes need to be devised.

5.4 Infrared Emissivity

Unlike in the microwave region, the infrared emissivity of sea water is very close to unity, and is very insensitive to surface temperature and roughness. There is an angular dependence which becomes significant at emission angles of 60° and greater but this would only be relevant for the extreme edges of the AVHRR scan. About 100 pixels at each end of the scan have zenith angle >60°. The tilting of facets of the surface by waves can increase the local zenith angle much beyond the mean value, but simulations of this effect have shown it to be negligible for cases where the mean local zenith angle is 60° or less.

5.5 The Skin Effect

The loss of heat by the ocean to the atmosphere results in a strong vertical temperature gradient in the top millimetre or so of the ocean. Since the skin depth for thermal infrared radiation is smaller than this, the radiation temperature derived from satellite observations is generally colder than that derived by an in situ measurement. The size of this 'skin effect' is controlled by local conditions such as rate of heat loss, wind speed, wave field, overlying cloud cover; and hence is very variable. Table 4 (adapted from Robinson, Wells and Charnock, 1983) shows the size of the skin effect in a number of published investigations. In the open ocean the effect can be to reduce the radiation temperature by several tenths of a degree Kelvin, although much bigger values have been seen.

For applications requiring satellite SST measurments as an equivalent 'bulk' SST, the skin effect will introduce significant bias errors. With the prospect of improved absolute accuracy of satellite measurements of SST, it is necessary to achieve an improved understanding of the skin effect, and, if possible, an improved parameterisation model (Saunders, 1973).

5.6 Diurnal Thermocline

In conditions of high insolation and minimal surface mixing (ie low wind) the heat gained by the ocean during the daylight hours can remain concentrated in a stable surface layer. The strong vertical thermal gradients can give rise to temperature differences of up to about 2 K in the top metres. Such features have been observed in the open ocean (eg Ostapoff and Worthem, 1974) and are strong functions of time of day, being strongest in late afternoon, and the integrated insolation and surface mixing conditions during that day. Since the diurnal thermocline is removed at night by mixing driven by heat loss to the atmosphere, its consequence on the interpretation of satellite SST values as bulk temperature measurements is very much reduced for satellites in a morning orbit. For satellite measurements taken in late afternoon it could be a significant source of error as the observed warming of the surface layer is highly transient and spatially variable.

6 FUTURE DEVELOPMENTS

6.1 Infrared Measurements

The launch in 1987 of ERS-1 (the first ESA Remote Sensing Satellite) is the main foreseeable development in SST measurement from satellites. In addition to a suite of radars (see for example Duchossois, 1983; Gillett, 1983), ERS-1 will carry the experimental Along Track Scanning Radiometer (ATSR), designed for the accurate measurement of SST. ATSR uses a simple scanning mechanism whereby the field of view sweeps out a cone inclined along the direction of the satellite flight path (Figure 10). In one revolution of the mirror the detectors receive radiation from two curved scan lines on the sea surface, along which the local zenith angle varies from 0° at nadir to 55° at the centre of the inclined scan. The two scan lines are

TABLE 4
VALUES OF THE SKIN EFFECT

a/LABORATORY MEASUREMENTS

Source	ΔT/K	Comments
Paulson & Parker (1972)	- 1.14 to - 1.81	Conditions of short fetch negligible wave generation. Barnes PRT-5 radiometer and mercury thermometer used.
Hill (1972)	~ - 1.0 ~ - 0.1	$u^* < 0.35\ ms^{-1}$ $u^* > 0.35\ ms^{-1}$. Change of roughness regime at u* 0.35 ms^{-1} due to onset of waves; short fetch. Radiometer and submerged turbulent jet.
Katsaros (1977)	- 0.4 to - 2.4	Barnes PRT-5 and disruption of surface layer; also depth soundings with 25μm thick resistance wire. No wind.

b/FIELD MEASUREMENTS

Source	ΔT/K	Comments
Woodcock & Stommel (1947)	- 0.5 to - 1.0	At night on a calm pond. Special mercury in glass thermometer.
Ewing & McAlister (1960)	- 0.6	Winds $< 0.5\ ms^{-1}$. No solar flux (at night). Radiometer and themistors, and also radiometer with surface disruption by pump.
Grassl (1976)	- 0.17 to - 0.21	Values obtained during GATE Wind speeds range 1-10 ms^{-1} PRT-5 and resistance thermometer.
Katsaros (1977)	up to - 3.4	Shielded so that $\overline{u}_{10} < 1\ ms^{-1}$. No waves. PRT-5 and disruption.

(Table 4 continued)

Schooley (1977)	- 0.2 ~ 0.0	Wind ~ 2.5 ms^{-1} - No cloud. Under cloud shadow. Depth sounding using thermopiles.
Nicholls (1979)	~ - 0.5 ± 0.3	Measured during JASIN; T_{bulk} measured by thermometer on board ships, $T_{surface}$ measured by PRT-4 airbourne radiometer. Atmospheric effects accounted for.
Simpson & Paulson (1980)	- 0.15 - 0.30	u ~ 5.5-9.2ms^{-1}. PRT-5 radiometer; T_{bulk} inferred as max values at temperature-wave peaks

(after Robinson, Wells and Charnock, 1983)

TABLE 5

SIMULATED ACCURACY ESTIMATES FOR AVHRR/2 AND ATSR
(Low Radiometer Noise : NEΔT 0.02 K - By Spatial Averaging)

Sensor	Single Channel (11 μm)		Day / Night Two-Channel (11 & 12 μm)		Night only Three-Channel (3.7, 11 & 12 μm)
	Clear	Hazy	Clear	Hazy	Clear*
AVHRR/2	-	-	0.38 K	0.55 K	0.15 K
ATSR	0.41 K	0.55 K	0.23 K	0.36 K	0.08 K

*for hazy conditions the 3.7 μm channel should not be used so the daytime values will apply.
(After Llewellyn-Jones et al, 1983)

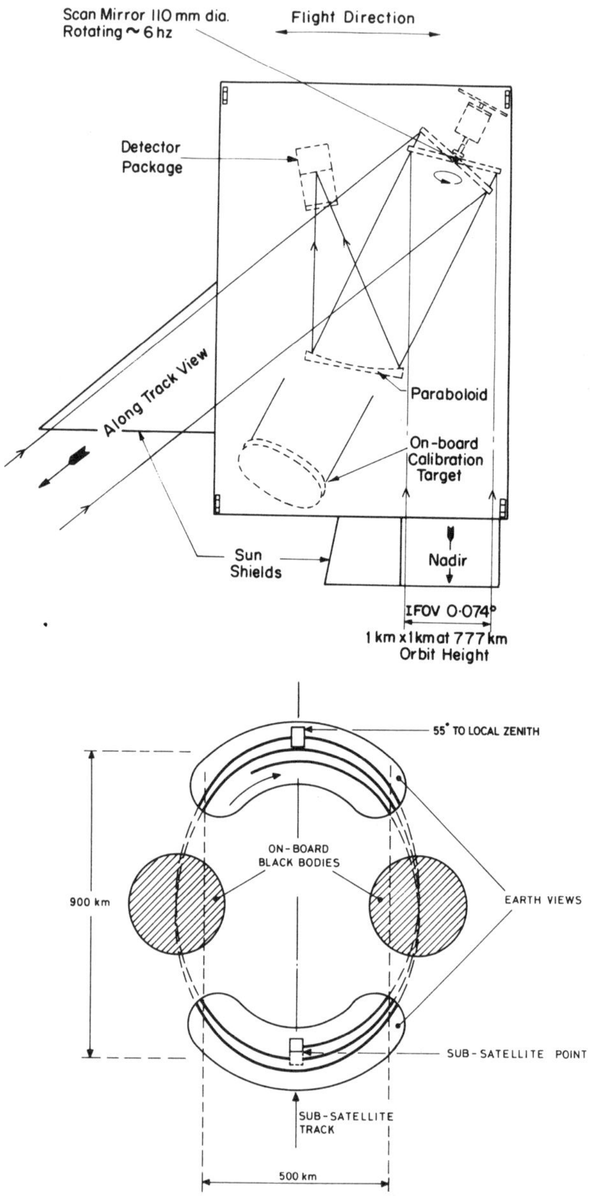

(a)

(b)

FIGURE 10
Schematic diagram of the Along Track Scanning Radiometer (ATSR) for the ERS-1 satellite (a). An inclined scan mirror sweeps out curved swaths from the nadir and an 'along track view', and includes measurements from two on-board calibration targets (b). The spacecraft motion overlays elements from the two swaths providing measurements of the same target through two atmospheric paths of differing lengths.

separated at their outer extremities while the detectors view two on-board calibration targets. The spacecraft motion overlays elements of the swaths after an interval of less than 140 secs. Assuming the characteristics of the atmosphere and ocean to be constant in this interval, two measurements of the ocean surface will have been made through two atmospheric paths of differing lengths. This is a direct measurement of the effect of the clear atmosphere on the radiation from the sea surface.

ATSR has three infrared channels in the same atmospheric windows as AVHRR/2 (3.7, 11 and 12 μm) to facilitate a comparison with the multichannel technique, and will cover a relatively narrow contiguous swath of 500 km with a sub-satellite spatial resolution of 1 km^2. Two nadir viewing microwave channels (23.8 and 36.5 GHz) are also included to provide an accurate determination of the total vertical water vapour amount. This not only provides the 'wet' tropospheric range correction for the ERS-1 Radar Altimeter, but improves the versatility of the infrared channels and provides continuous self-consistency checks. Furthermore the measurement of the atmospheric water content is of itself relevant to climate studies.

The performance of ATSR has been simulated by Zavody (1982) using an atmospheric radiation transfer model. The results are summarised in Table 5. ATSR will provide measurements of SST, on a 50 km grid, with absolute accuracy of better than 0.5 K, in conditions of up to 80% cloud cover, assuming improvements in current cloud clearing techniques. The simulated accuracy is shown in relation to the WCRP requirements in Figure 1.

The double angle measurement of ATSR provides an increased resistance to contamination of the SST measurement by aerosols. The conventional multichannel SST retrieval can be invalidated by aerosols in the stratosphere, injected by volcanic eruptions, as well as those in the troposphere.

6.2 Microwave Measurements

Scanning microwave radiometers are attractive for climate studies in being able to measure SST through even thick cloud cover. To realise this potential, however, improved instruments will be needed with, for example, larger antennae, and it is doubtful whether this is achievable within the period of the WCRP.

7 CONCLUSIONS AND RECOMMENDATIONS

The main conclusions and recommendations can be summarised :

a/ The accuracies achieved with current infrared radiometers (AVHRR/2) have approached the levels required by the WCRP for large scale studies. While some reduction in the uncertainties may be achieved by averaging many measurements, new measurement techniques will be needed to bring the absolute uncertainties within the required ranges. Multi-angle scanning, such as employed by ATSR, is such a development.

b/ Multichannel SST retrieval algorithms may be improved by better knowledge of global distributions of the atmospheric temperature and water vapour profiles. Suitable measurements are scarce in the Southern Hemisphere, and accurate satellite soundings could provide a method for improvement. Careful consideration should be given to advanced instrumentation to provide improved atmospheric soundings. The retrieval algorithms could also be improved by better knowledge of the spectral properties of the atmospheric molecules in the atmospheric 'windows'.

c/ Incorporation of an airmass dependence in the SST retrieval algorithms would produce an improvement in accuracy and coverage in current SST measurements.

d/ 'Split-window' measurements (11, 12 μm) can be made when sun glitter renders the 3.7 μm channel unusable. This can be a large proportion of the swath.

e/ Much remains to be done in developing computationally efficient cloud clearing algorithms. The results presented here do not exhaust all possible combinations of techniques currently available, but they highlight some of the problems to be overcome.

f/ Attention has been focussed on the capabilities and problems of infrared instruments on polar-orbiting satellites. For some purposes the coverage of this type of satellite may be inadequate and the high temporal resolution offered by geostationary satellites would be an advantage. SST measurement capability, operating with 'split-window' channels, should be part of future geostationary payloads.

REFERENCES

Arking A and Childs J D, 1983, Extraction of cloud cover parameters from multispectral satellite measurements. Submitted to J. Climate and Appl. Met.

Barnett T P, 1981, Statistical prediction of North America air temperatures from Pacific predictors. Mon. Wea. Rev. 109, pp 1021-1041.

Bernstein R L, 1982a. SST mapping with the Seasat Microwave radiometer. J. Geophys. Res. 87, pp 7865-7872.

Bernstein R L, 1982b, Sea surface temperature estimation using the NOAA-6 Satellite Advanced Very High Resolution Radiometer. J. Geophys. Res. 87, pp 9455-9465.

Coakley J A and Bretherton F P, 1982, Cloud cover from high resolution scanner data : detecting and allowing for partially filled fields of view. J. Geophys. Res. 87, pp 4917-4932.

Cox C S and Munk W H, 1954, Statistics of the sea surface derived from sun glitter. J. Mar. Res. 13, pp 198-227

Crosby D S and Glasser K S, 1978, Radiance estimates from truncated observations. J. Appl. Met. 17, pp 1712-1715.

Deschamps P Y and Phulpin T, 1980, Atmospheric correction of infrared measurements of sea surface temperature using channels at 3.7 11 and 12 μm. Bound. Layer. Met. 18, pp 131-143.

Dozier J, 1981, A method for satellite identification of surface temperature fields of subpixel resolution. Remote Sensing of Environment, 11, pp 221-119.

Duchossois G, 1983, Status and future plans for ERS-1. In 'Satellite microwave remote sensing' by T D Allan (ed). Ellis Horwood, pp 501-514.

Ewing G and McAlister E D, 1960, On the thermal boundary layer of ocean. Science, 131, pp 1374-1376.

Gillett P R C, 1983, ERS-1 : An ice and ocean monitoring system. J. Brit. Interplanetary Soc. 36, pp 387-393

Grassl H, 1976, The dependence of the measured cool skin of the ocean on wind stress and total heat flux. Bound. Layer Met. 10, pp 465-474.

Hill R H, 1972, Laboratory measurement of heat transfer and thermal structure near an air-water interface. J. Phys. Oceanog. 2, pp 190-198.

Hoskins B J and D K Karoly, 1981, The steady linear response of a spherical atmosphere to thermal and orographic forcing. J. Atmos. Sci. 38, pp 1179-1196.

Houghton J T and Morel P, 1983, The world climate research programme in 'The Global Climate' by J T Houghton (ed). Cambridge University Press. In the Press.

Katsaros K B, 1977, The sea surface temperature deviation at very low wind speeds, is there a limit? Tellus, 29, pp 220-239.

Lauritson L, Nelson G J and Porto F W, 1979, Data extraction and calibration of TIROS-N/NOAA radiometers. NOAA Tech. Memo. NESS 107 44-46.

Llewellyn-Jones D T, Minnett P J, Saunders R W and Zavody A M, 1983, Satellite multichannel infrared measurements of sea-surface temperature of the NE Atlantic Ocean using AVHRR/2. Submitted to Quart. J. R. Met. Soc.

Madden R A and Ramanathan V, 1980, Detecting climate change due to increasing carbon dioxide. Science, 109, pp 763-768.

McClain E P, Pichel W, Walton C, Ahmad Z and Sutton J, 1983, Multichannel improvements to satellite-derived global sea surface temperatures. Advances in Space Research, 2, pp 43-47.

McClain E P and Walton C W, 1982, Satellite zenith angle corrections for satellite measurements. Unpublished manuscript.

McMillin L M, 1975, Estimation of sea surface temperature from two infrared window measurements with different absorption. J. Geophys. Res. 80, pp 5113-5117.

Nicholls S, 1979, Intercomparisons between the MRF Hercules and surface observations from the JASIN meteorological ships. JASIN News No. 19, pp 9-11. Unpublished manuscript.

Njoku E, 1982, Passive microwave remote sensing of the earth from space a review. Proc IEEE, 70, pp 728-750.

Ostapoff F and Worthem S, 1974, The intradiurnal temperature variation in the upper ocean layer. J. Phys. Oceanog. 4, pp 601-612.

Paulson C A and Simpson J J, 1981, The temperature difference across the cool skin of the ocean. J. Geophys. Res. 86, pp 11044-11054.

Prabhakara C, Dalu G and Kunde V G, 1974, Estimation of sea surface temperature from remote sensing in the 11 to 13 μm window region. J. Geophys. Res. 79, pp 5039-5044.

Robinson I S, Wells N and Charnock H, 1983, The sea surface thermal boundary layer and its relevance to the measurement of sea surface temperature by airborne and spaceborne radiometers. Intl. J. Remote Sensing. In the Press.

Saunders P M, 1973, The skin temperature of the ocean, a review. Mem. Soc. Roy. Sci. de Liege, 6 ser., pp 93-98.

Schooley A H, 1977, The temperature of the ocean skin related to cloud shadows. J. Phys. Oceanog. 7. pp 486-487.

Taylor P K, 1983, The scanning multichannel microwave radiometer - an assessment. In 'Satellite microwave remote sensing', by T D Allan (ed). Ellis Horwood. pp 463-480.

Wilheit T T and Chang A T C, 1980, An algorithm for retrieval of ocean surface and atmospheric parameters from observations of the scanning multichannel microwave radiometer. Radio Sc. 15, pp 525-544.

Woodcock A H and Stommel H, 1947, Temperatures observed near the surface of a fresh water pond at night. J. Meteorol. 4, pp 102-103.

World Climate Research Programme, 1981, Report of meeting on co-ordination of plans for future satellite observing systems and ocean experiments to be organised within the WCRP, Jan. 1981. WCP-8.

Zavody A M, 1982, Appendix J of 'The Along Track Scanning Radiometer with Microwave Sounder'. Proposal submitted to ESA in response to ERS-1 Announcement of Opportunity. Rutherford Appleton Laboratory. Unpublished manuscript.

SATELLITE SEA SURFACE TEMPERATURE DETERMINATION FROM MICROWAVE AND INFRARED RADIOMETRY

R.L. Bernstein

California Space Institute, A-021
Scripps Institution of Oceanography
La Jolla, CA 92093 USA

ABSTRACT

Starting in 1978, the present generation of satellite-borne microwave and infrared radiometers are beginning to demonstrate sea surface temperature (SST) estimation accuracies of 0.5-1.0° C, sufficient to map on a global scale the large-scale SST field and its anomaly with respect to climatological norm. In addition, these sensors provide data on the wind at the sea surface, vertically integrated water vapor and liquid water, rain rate, and atmospheric aerosols, all of which are quite important for climate research. In some cases and for certain satellite instruments, evidence is accumulating which suggests that the 0.5°C rms disagreement between ship or buoy data, and that derived from satellite, may actually be an upper bound on satellite accuracy. This is a result of imperfect space-time co-location, as well as true differences between the 'skin' temperature measured by radiometry, and the bulk thermometric temperature usually obtained from ships and buoys. Thus the possibility begins to emerge for determining from space the skin temperature relative to bulk thermometric temperature, at points where the latter is determined, for example, from simple drifting buoys. If this difference can be determined to a few tenths of a degree Celcius, it can be related to the heat and momentum fluxes across the air-sea interface. Since momentum flux can be separately determined by active and passive microwave techniques, it may thus prove possible to determine heat flux (long-wave radiative, plus latent and sensible components) to useful accuracy. This, combined with geostationary satellite methods for measuring the short wave radiation reaching the surface, could provide the means for

C. Gautier and M. Fieux (eds.), Large-Scale Oceanographic Experiments and Satellites, 87–98.

determining the total heat exchange between the ocean and atmosphere, globally and over long periods of time, on a practical, cost-effective basis, with significant implications for climate studies.

1. INTRODUCTION

The state of the art in radiometric determination of sea surface temperature from satellite-based instruments has advanced significantly in recent years, reaching the point where the resultant data may be used in an increasing number of applications. This paper is intended as a brief review of that progress, illustrated with examples, discussion of accuracies thus far achieved, and concluding with some admittedly speculative remarks extrapolating the recent trend of advances into the future, five to ten years from now.

Radiometer estimates of sea surface temperature are based on the measured brightness temperature of the surface, which in turn depends on the emissivity of the surface, as well as the actual temperature of the surface. The definition of "surface temperature" becomes itself critical, especially as the level of accuracy improves to the point where true variations in temperature within the upper few meters of water become comparable in magnitude. Very recent work by Gasparovic (1983) provides an excellent discussion of the problem, illustrating how microwave radiation is emitted from the upper centimeter of water, while infrared radiation is from the upper 10 to 100 μm of water. Since true variations in temperature on the order of 0.5°C occur across this domain (Saunders, 1967; Katsaros, 1980; Paulson and Simpson, 1981), some care must be taken when comparing radiometric estimates of sea surface temperature with traditional measurements from buckets, buoys, XBT's and the like, which variously come from the upper 1 to 10 meters.

2. INFRARED MEASUREMENTS

The best satellite estimates of sea surface temperature presently are derived using the infrared multi-spectral measurements of brightness temperature from the AVHRR (Advanced Very High Resolution Radiometer) of the Tiros-N series of NOAA, polar-orbiting weather satellites. The AVHRR is a five channel (0.65, 0.9, 3.7, 11, and 12 μm) sensor, able to resolve brightness temperature at 1 km resolution and 0.1 K accuracy. Procedures for linearly combining temperatures from the 3.7, 11, and 12 μm channels were described by Deschamps and Phulpin (1980), and McClain (1981), and recent results of applying these or related techniques (Bernstein, 1982a; McClain et al., 1982) are yielding agreement at the 0.6 K rms level with in-water observations from buoys and XBT's. The degree to which satellite and in-water measurements are spatially and temporally coincident then begins to be an important factor, as does the previously noted difference in temperature within the upper few meters of water.

It is at least possible, therefore, that AVHRR-derived estimates of the temperature of the surface (upper 10 to 100 μm) are accurate to levels significantly better than 0.6 K. Experimental work to determine whether this is so will require very near surface radiometric measurements of high accuracy, precisely co-located in time and space with the AVHRR measurements. Low-flying aircraft carrying radiometers provide one means of making such measurements, and results by Peacock and Gasparovic (1983) at the 0.3 K agreement level are encouraging.

The primary atmospheric error sources for sea surface temperature estimation from AVHRR are clouds, water vapor, and aerosols. The existing procedures assume, or must somehow insure, that the brightness temperatures are for cloud-free fields of view. This may be called the cloud clearing problem, and is addressed by two different approaches.

One approach uses daytime data, where visible/near-infrared (reflected solar) radiance information is available from the first two channels of the AVHRR. Those fields-of-view which are even partially cloud-contaminated have higher radiance levels, compared with those viewing the dark ocean background. This approach has two drawbacks: sun glint, and variable amounts of aerosol in the cloud-free atmosphere. Sun glint destroys the dark background, while changes in the general aerosol load of the atmosphere creates variable solar backscattering radiances from the atmosphere itself. Despite these limitations this approach has been used successfully (Bernstein, 1982a) to obtain 0.6 K agreement between NOAA-6 AVHRR and nearby XBT and buoy data.

The second approach uses night-time AVHRR data, where only the three thermal infrared channels are available. As described by McClain et al. (1982), the local variability in 11 μm brightness temperature is computed over a small array of adjacent fields-of-view. If the variability significantly exceeds the sensor noise level, then one or more of the fields-of-view are assumed to be cloud-contaminated. In order to eliminate situations of uniform temperature solid cloud cover such as low stratus, the difference in brightness temperature at 3.7 and 11 μm is examined. Since cloud emissivity is quite different at these two wavelengths, stratus cloud will produce a unique difference in temperature. This approach has also yielded agreement at the 0.6 K level with in-water measurements.

The latter approach has potential for eliminating some of the errors induced by variable amounts of aerosol, since aerosol absorption is quite different at 3.7 and 11 μm. Recent work by Walton (1983) and others suggests that techniques based on this factor may at least reduce the large errors of several degrees associated with volcanic dust such as that produced by the El Chichon explosion of 1982.

A second instrument on the Tiros-N series spacecraft, the HIRS/2, has vertical profiling of atmospheric temperature and moisture as its principal objective. Yet new work by Susskind et al., (1982) shows some promise for sea surface temperature estimation. The HIRS/2 instrument has about 30 km spatial resolution, but observes at 19 different spectral channels. The Susskind procedure groups these fields-of-view into 125 km sized arrays, and early, quite preliminary reports suggest 1.0 to 1.5 K agreement with in-water data.

3. MICROWAVE MEASUREMENTS

Turning now to microwave, only one type of instrument with sea surface temperature capability has been launched. The five channel (6.6, 10.7, 18, 21, and 37 gHz) SMMR (Scanning Multi-frequency, Microwave Radiometer) operated for a few months on Seasat, and has also continued to operate for five years on NIMBUS-7. At 6.6 gHz even the cloudy atmosphere is quite transparent, permitting measurement of surface brightness temperatures with only small corrections needed for water vapor and liquid water, which are provided by the observations at the three highest frequencies. Unfortunately, the emissivity of the ocean at 6.6 gHz is about 0.5, and varies strongly with wind speed and sea state. The 10.7 gHz frequency is less sensitive to sea surface temperature, and more so to sea state, and may thus be used along with the higher and lower frequencies to estimate both sea surface temperature and wind speed.

The SMMR data is potentially quite powerful for large-scale climatic purposes. Its principal limitations are spatial resolution (150 km) and engineering factors related to compromises made in producing this first generation instrument. Nevertheless, it has been possible to derive monthly mean sea surface temperature maps having about 1.0 K agreement with the routinely reported shipboard measurements (Bernstein, 1982b; Bernstein and Morris, 1983).

4. APPLICATION TO THE EQUATORIAL PACIFIC

The ability of the data from the various infrared and microwave radiometers to dramatically improve our knowledge of sea surface temperature, relative to shipboard techniques, is best demonstrated in the eastern tropical Pacific, where ship traffic is light. Beginning with a climatological field, such as that of Levitus and Oort (1977), one can see the well-known tongue of colder water extending westward from South America along the equator (Figure 1). Within the boxed area of Figure 1

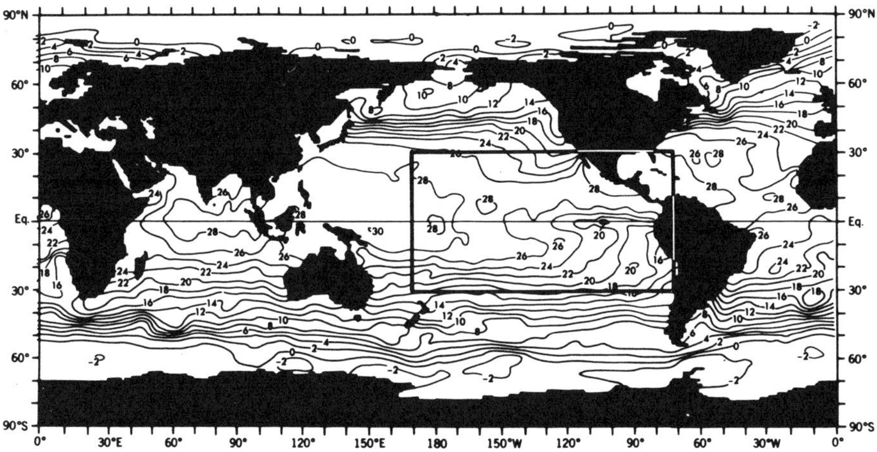

Fig. 1 Climatological sea surface temperature distribution for September, eastern North Pacific. Box denotes subarea covered by Figure 2 (from Levitus and Oort, 1977).

the routinely available ship data (Figure 2) leaves many data void areas, although careful semi-monthly mapping attempts (Liu, 1982) can reveal the cold tongue and other large-scale structures. A one week composite of SMMR data (Figure 3), for the same time period and within the boxed area of Figure 2, shows 1000 km wave-like structures along the temperature gradient north of and parallel to the equator. These waves, first discovered by Legeckis (1977), appear to travel west at about 40 km/day, and in fact week-to-week SMMR map sequences

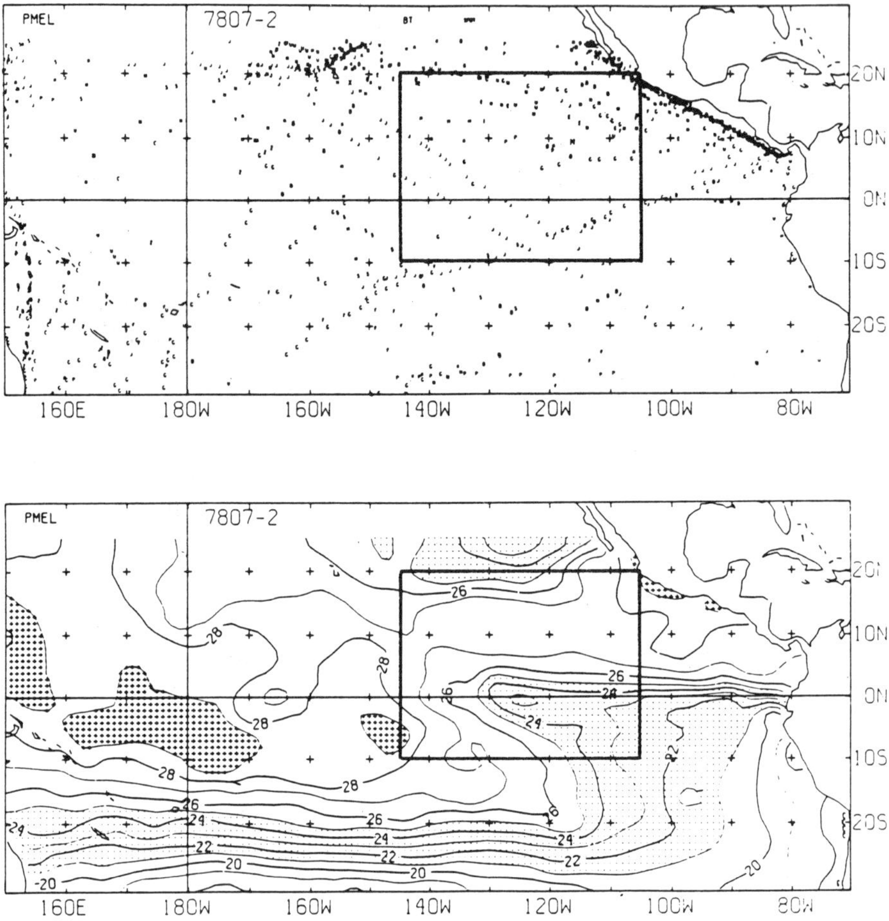

Fig. 2 Distribution of observations and map of sea surface temperature, for all ship reports in the eastern North Pacific, 16-31 July 1978. Box denotes subarea covered by Figure 3 during the same time period (from Liu, 1982).

define similar behavior. Finally, within the boxed area of Figure 3, AVHRR data at 1 km resolution (Figure 4) determines the spatial structure to a remarkable degree, and the absolute value of sea surface temperature to within about 0.5 K of shipboard measurements (Figure 5). The absolute sea surface temperature estimates were derived employing the techniques described in Bernstein (1982a).

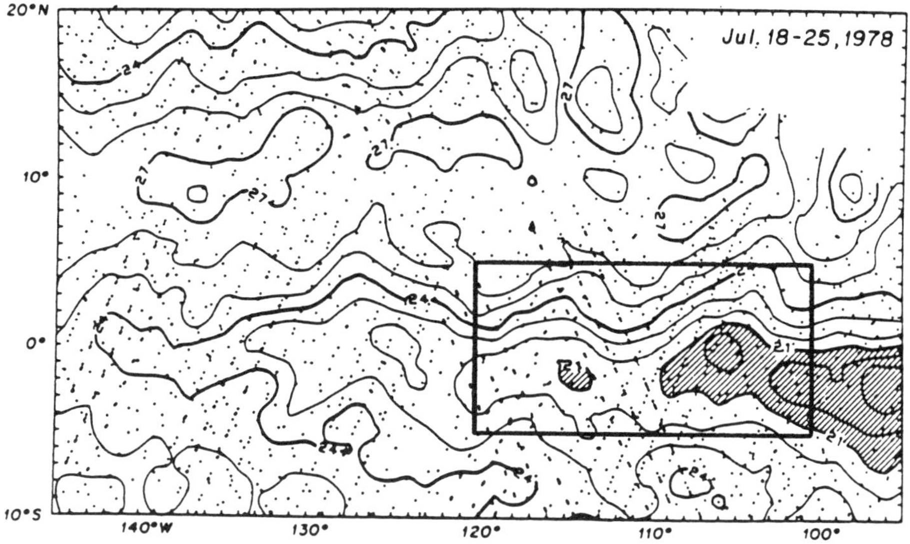

Fig. 3 Mapping and distribution of Seasat SMMR observations of sea surface temperature, for 18-25 July 1978 in the eastern Tropical Pacific. Box denotes subarea covered by Figure 4 during same season of a different year. (From Bernstein and Morris, 1983).

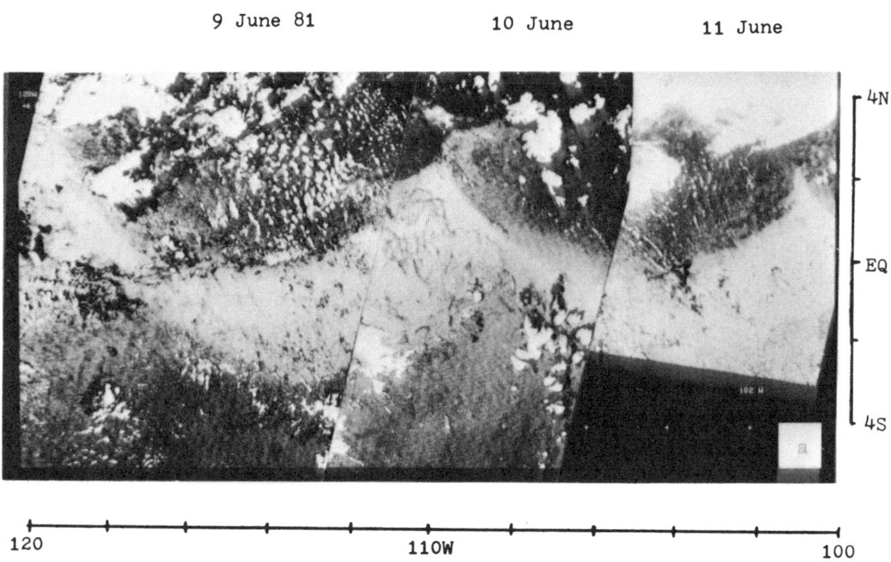

Fig. 4 Composite of three AVHRR 3.7 µm brightness temperature images for June 9, 10 and 11, 1981 in eastern Tropical Pacific.

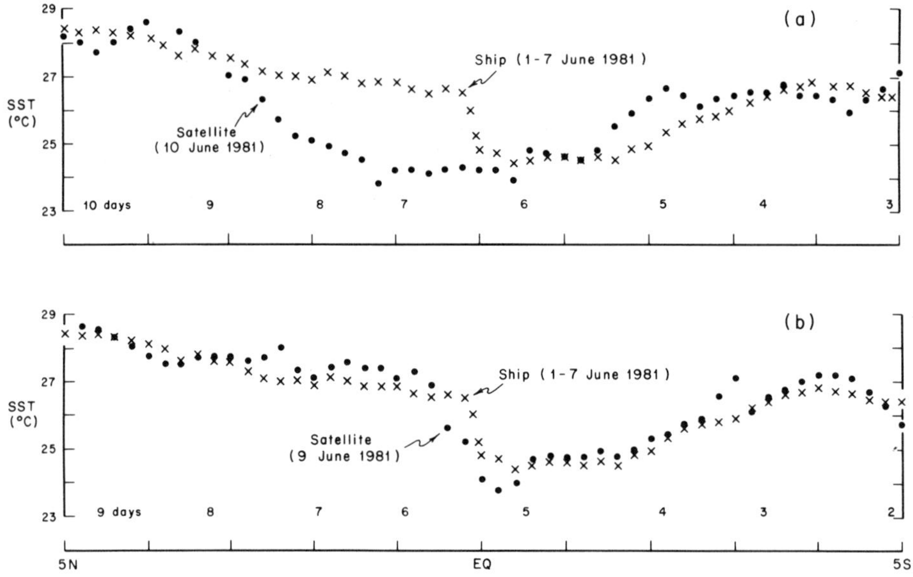

Fig. 5 (top): Comparison of AVHRR-derived temperatures (from Figure 4) along 110°W with shipboard measurements provided by P. Pullen (PMEL); (bottom): same as top except adjusted for effects of westward propagating wave and time difference between ship and satellite observations.

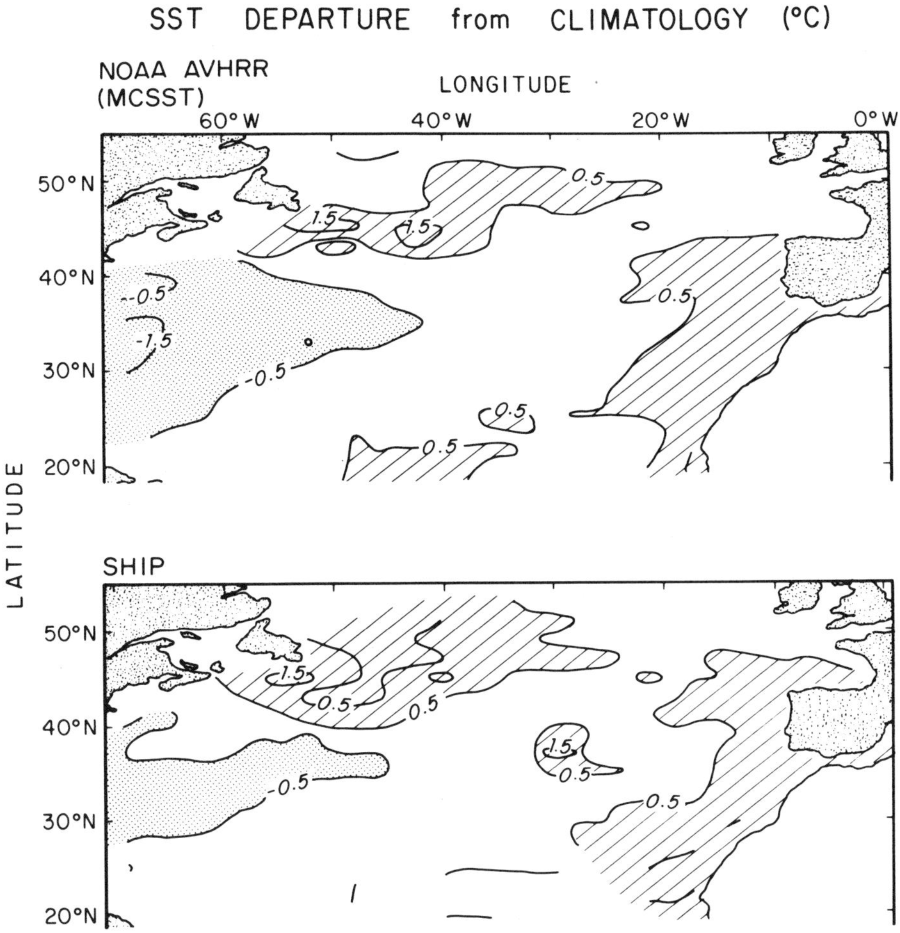

Fig. 6 (top): North Atlantic sea surface temperature anomaly maps for December 1981, derived from NOAA AVHRR (multichannel sea surface temperature) data; (bottom): the same except from routinely reported ship observations. (Charts redrawn from materials provided by NASA Jet Propulsion Laboratory).

5. CONCLUSION

Having reviewed some evidence that the capability of satellite radiometry for sea surface temperature has reached the level of useful accuracy, the question arises as to how this capability will in fact be used. There is little doubt that satellite-based maps of sea surface temperature anomaly (deviation from climatological norm) will be increasingly applied by those who have until now relied on ship-based maps. Thus, for example, anomaly charts such as Figure 6 for the North Atlantic have been produced at the NASA Jet Propulsion Laboratory, using NOAA-AVHRR operationally derived sea surface temperatures. The pattern of anomaly is in very good agreement with that based on routinely collected ship data. It is also quite likely that the role of high resolution sea surface temperature maps and imagery will become an essential one for many field experiments, as has been the case with such U.S. field programs as FRONTS 80, CODE, and Warm Rings.

As far as the future is concerned, the possible increase in demonstrated accuracy for deriving skin temperature, to levels of a few tenth of a degree Kelvin, may open the way to fundamental advances in air-sea interaction. As noted in some of the papers cited earlier, the actual surface temperature deviates from the temperature at 1 cm or deeper, by an amount thought to be proportional to the heat flux (sensible, latent, and radiative), and inversely to the friction velocity, across the air-sea interface. It is thus at least conceivable that this temperature difference may be determined by simultaneous infrared and microwave radiometric measurements from space, while the friction velocity is determined from either active or passive microwave techniques. While still extremely speculative, the possibility thus exists for directly monitoring the heat flux, a parameter of critical importance to climate study, and presently the most difficult to measure. Even if any of a number of obstacles to accomplishing this prove insurmountable, it is likely that much will be learned along the way about the processes at work across the air-sea interface.

REFERENCES

Bernstein, R.L., 1982a, Sea surface temperature estimation using the NOAA-6 satellite Advanced Very High Resolution Radiometer. J. Geophys. Res., 87, 9455-9465.

Bernstein, R.L., 1982b, Sea surface temperature mapping with the Seasat microwave radiometer, J. Geophys. Res., 87, 7865-7872.

Bernstein, R.L., and J.H. Morris, 1983, Tropical and mid-latitude North Pacific sea surface temperature variability from the Seasat SMMR, J. Geophys. Res., 88, 1877-1891.

Deschamps, P.Y., and T. Phulpin, 1980, Atmospheric corrections of infrared measurements of sea surface temperature using channels at 3.7, 11, and 12 um, Boundary Layer Meteorol., 18, 131-143.

Gasparovic, R.F., 1983, "Effects of spatial variability on remotely sensed sea surface temperatures." Report STD-R-822, John Hopkins Applied Physics Laboratory, Laurel, Maryland.

Katsaros, K., 1980, Radiative sensing of sea surface temperature, in Air-Sea Interaction: Instruments and Methods, edited by F. Dobson, L. Hasse, and R. Davis, 801, Plenum, New York.

Legeckis, R.V., 1977, Long waves in the eastern equatorial Pacific Ocean: a view from a geostationary satellite, Science, 197, 1179-1181.

Levitus, S. and A.H. Oort, 1977, Global analysis of oceanographic data, Bull. Am. Meteorol. Soc., 58, 1270-1284.

Liu, Cho-Teng, 1982, Analysis of tropical Pacific sea surface temperatures for 1975 to 1980. NOAA Tech. Mem. ERL PMEL-34, Pacific Marine Environmental Laboratory, Seattle, Washington, 160.

McClain, E.P., 1981, Multiple atmospheric-window techniques for satellite-derived sea surface temperatures, in Oceanography from Space. Edited by J.F.R. Gower, Plenum, New York, 73-85.

McClain, E.P., W.G. Pickel, C.C. Walton, Z. Ahmed, and J. Sutton, 1982, Multichannel improvements to satellite-derived global sea surface temperatures. XXIV COSPAR Meeting, Ottawa, May 1982.

Paulson, C.A., and J.J. Simpson, 1981, The temperature difference across the cool skin of the ocean. _J. Geophys. Res._, _86_, 11,044-11,054.

Peacock, K. and R.F. Gasparovic, 1983, Aircraft validation of multichannel SST algorithms for NOAA-7 AVHRR imagery, _Trans. Am. Geophys. Un._, _64_, pp. 232.

Saunders, P.M., 1967, The temperature at the ocean-air interface, _J. Atmos. Sci._, _24_, 269-273.

Susskind, J., J. Rosenfield, D. Renter, and M.T. Chahine, 1982, The GLAS physical inversion method for analysis of HIRS 2/MSU sounding data. NASA Tech. Mem. 84936, Goddard Space Flight Center, Greenbelt, Maryland, pp. 101.

Walton, C.C., 1983, Aerosol effects on satellite multi-channel sea surface temperature measurements. _Trans. Am. Geophys. Un._, _64_, pp. 232.

OCEAN SURFACE WIND STRESS

D.E. Harrison

Center for Meteorology and Physical Oceanography,
M.I.T., Cambridge, MA 02139 USA

ABSTRACT

There is an urgent need in physical oceanography, meteorology, marine operations and climate research for improved surface wind and wind stress information. Much can be done to improve our knowledge through better collection of existing observations, deployment of properly instrumented drifting buoys, better calibration and collection of cloud motion vectors, increased scrutiny and development of analysis/forecast methods of surface data assimilation and more widespread direct measurement of surface stress. But the space and time sampling constraints on such techniques will inevitably limit the improvement to less than is desireable or needed. Only satellite-borne sensors, adequately deployed and calibrated against high quality surface data can provide the needed information. The microwave scatterometer shows promise for meeting the required accuracy needs, and may soon revolutionize our knowledge of the surface wind and stress fields.

A. Introduction

The wind stress at the ocean surface is physically important for many reasons. It directly or indirectly determines many of the currents in the sea, the heat capacity of the upper ocean (through its role in setting the mixed layer depth), the wave state of the sea surface, etc. Climate studies, ocean circulation studies, efficient maritime operations planning, weather prediction and other important activities require that the surface stress be known.

C. Gautier and M. Fieux (eds.), Large-Scale Oceanographic Experiments and Satellites, 99–115.

Unfortunately the surface stress is not as well known as one would like, either climatologically or on shorter time scales. First of all, very few direct measurements of stress have ever been made at sea. Only rather recently in fact, have the techniques of measuring stress been shown to give the same values to ~10% (1). Instead of stress data, it has been necessary to work with near surface wind data and to assume that the wind and the stress can be related through a bulk aerodynamic formula:

$$\vec{\tau} = \rho\, C_D\, |\vec{u}|\, \vec{u}$$

where ρ is the density of air, and C_D, the drag coefficient, is assumed to be a function only of wind speed ($|\vec{u}|$) and atmospheric boundary layer stability (T_{air}-SST). Typically $\vec{u}$ and T_{air} are assumed known at 10m or 19.5m above the sea surface.

Many different forms for the dependence of C_D upon windspeed and atmospheric stability appear in the literature. Most of the recent forms are referenced in (1), and (2) has provided tables of C_D for all plausible conditions for a form consistent with that preferred by (1). Figure 1 shows the acceptable envelope of values of open ocean C_D, for neutral stability, as a function of windspeed according to Large. Uncertainty concerning the form of C_D appears to have been diminished by this recent work; hopefully further studies under different environmental conditions will confirm these recent findings. Evidence exists, however, that shallow water values of C_D may be quite different from these deep water values. There is a pressing need to acquire a larger set of measurements of stress encompassing a wide range of environmental conditions, in order to determine convincingly the functional form of C_D.

Even if the form of C_D is known, the use of bulk formulae to relate near surface wind to surface wind stress imposes restrictions on the scales of stress that can be sampled. Because C_D is determined empirically, it is a "best fit" to actual stress data. The bulk formula thus can only be expected to provide a satisfactory stress value in general when applied to averages long enough in time to span a considerable range of atmospheric conditions. A bulk formula applied to hourly or even daily average wind data may lead to large (>100%) errors in stress. In middle latitudes averages over 3 to 5 days appear to be appropriate (1).

Fortunately for low frequency ocean circulation studies and climate studies, averages over such time periods are not a constraint. However, there remain many problems, e.g., surface

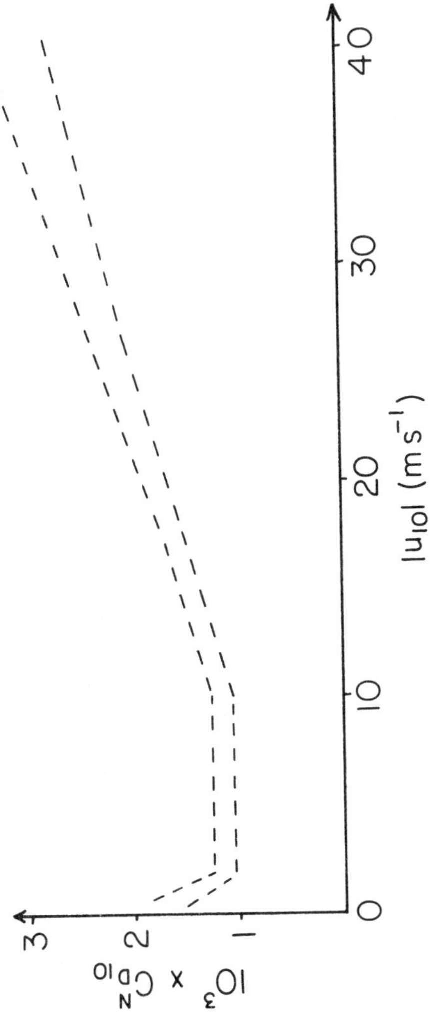

Figure 1. Range of acceptable values of neutral stability, 10m height drag coefficient as a function of wind speed, according to W. Large (personal communication, 1983), is indicated by area between dashed lines. (See text).

wave forecasting, where such averaging cannot be tolerated. Special studies will be required to extend the bulk formulae to be applicable at shorter time scales, and information beyond windspeed and air-sea temperature difference will most likely be required for such extension.

Because so much of the atmospheric variability is found at time scales of five days or less, sampling is a critical problem. Accurate low frequency averages can be obtained only if the higher frequency motions are adequately observed. The fractional amount of power in high frequencies is a strong function of location and of season. Figure 2 shows schematically estimates of power density spectra of east wind for northern middle-latitude conditions (based on geostrophic calculations, (3)) for Midway Island (~28°N), for Ponape Island (~7°N) and Canton Island (~1°S), for periods somewhat greater than 1 day and less than 180 days. Note the large differences in the fractional power contained between 1 day and 5 day periods at these different locations, but it is substantial in each case. There can also be substantial power at periods of 1 day or less in some locations.

The minimum sampling required for a given level of accuracy, as a function of averaging interval, has not been explored systematically, even though this is an important design criterion for wind programs. However, even in trade wind areas the required sampling can be quite large; recent calculations by the author and D. Luther show that roughly 20 perfect random observations per month are required to obtain a monthly mean zonal wind time series at Canton Island that correlates with the correct timeseries at >0.9 and has an RMS error of $\lesssim 0.5$ ms^{-1}.

Clearly even greater sampling is required poleward of the tropics (Figure 2); specific sampling criteria should be worked out in the near future.

A second source of concern is, of course, the accuracy of individual near surface wind observations. Wind data at present are available from ship reports and a few research buoys. Although not near surface, cloud motion vectors, based on cloud pattern movement at heights of 850-950 mb., and then empirically adjusted to nominal near surface values are the source of much interest at present. We shall consider each type of data in turn.

B. Data Sources

The overwhelming majority of near surface wind data is obtained from merchant ship reports. These reports are based

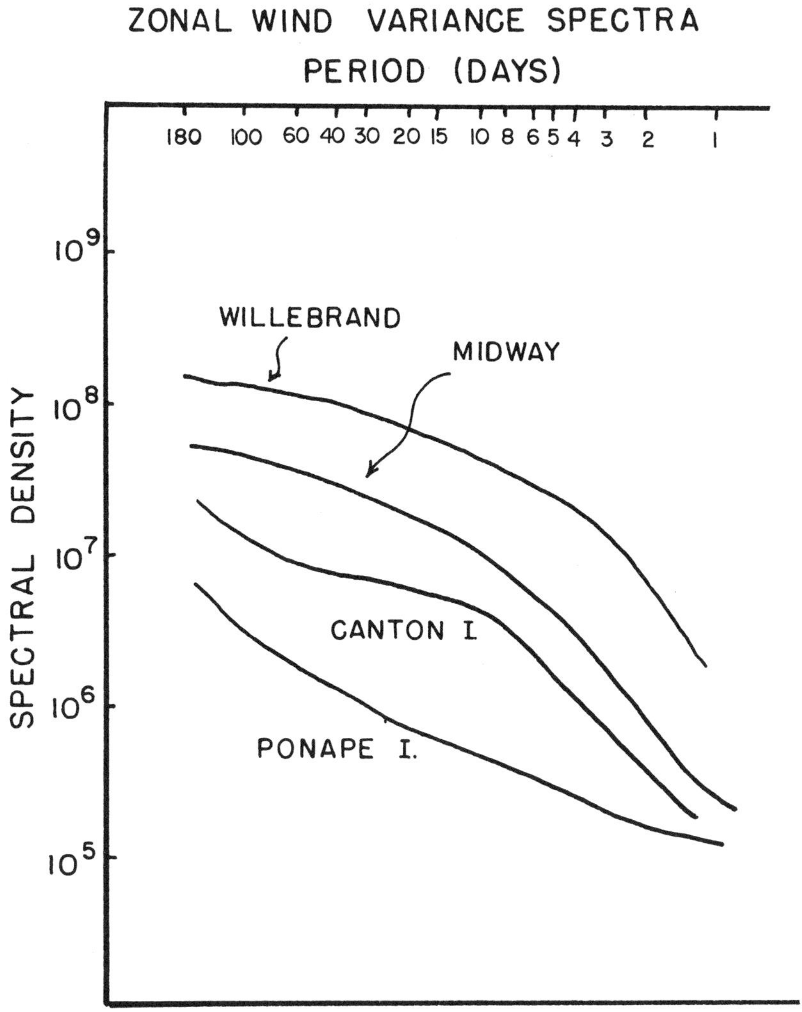

Figure 2. Schematic power spectral density of zonal wind at different locations. Note the significant amounts of power in the 1 to 5 day band even at tropical latitudes. Considerable sampling density is needed if aliasing of long-period changes is to be avoided. (See text).

either on Beaufort observations of the physical state of the sea surface (ripples, white caps, spray, etc.) or on anemometer readings. The Beaufort scale has an intrinsic uncertainty of ~2 ms^{-1}, before considering the judgement of an individual observer, and wind direction must be determined independently of windspeed. With enough observers one might hope that the errors would have a substantial random element. Anemometer readings cannot be assigned any convenient error estimate, because the method of averaging, flow interference by the ship superstructure (which is often strongly dependent on ship relative heading), anemometer height correction (or lack of correction) and uncertain maintenance and calibration each can introduce errors of more than 1 ms^{-1} into a given reading; note that these will not be random errors in most cases.

Research buoys use anemometer systems of different types, and are not subject to any standard performance specifications. Simple survival over many months remains a problem, so efforts to estimate drift or other types of sensor deterioration have not been typical. Further the anemometers have been deployed on a variety of buoys, leading to various sensor locations and very different types of buoy motion environments. Only limited intercomparison activities have been undertaken, not surprisingly, given the nearly "one-off" character of many systems. In any case, the total amount of data from buoys to date is small enough not to signficantly affect the global data base.

Cloud motion vectors, adjusted to nominal 10m height, have been the focus of much recent interest. At least two analysis procedures have been carried out in the U.S. NOAA/NESS has routinely provided CMVs from its geostationary satellite visual images. These vectors are produced objectively by a pattern-optimization procedure. The Space Sciences Engineering Center at U. Wisconsin has provided smaller quantities of CMVs by tracking individual cloud clusters. Differences between the results from the two analysis procedures exceeded 5 ms^{-1} ~30% of the time in one set of comparisons (4). The relationship between cloud motion and winds around the cloud is undocumented, as is the uncertainty introduced by shearing of the clouds by the wind field. Whatever the error sources, CMVs provide a sufficiently large data set of velocity information that the aliasing of high frequencies into lower frequencies is hopefully less of a problem than with the ship data. Some aliasing will occur because the clear sky and overcast sky are never sampled.

A major source of uncertainty in winds derived from CMVs is the need to adjust a given CMV or average CMV from "cloud height" to 10 m height (Figure 3). Typically the height is uncertain to ~50 mb, and the vertical shear of the horizontal wind between

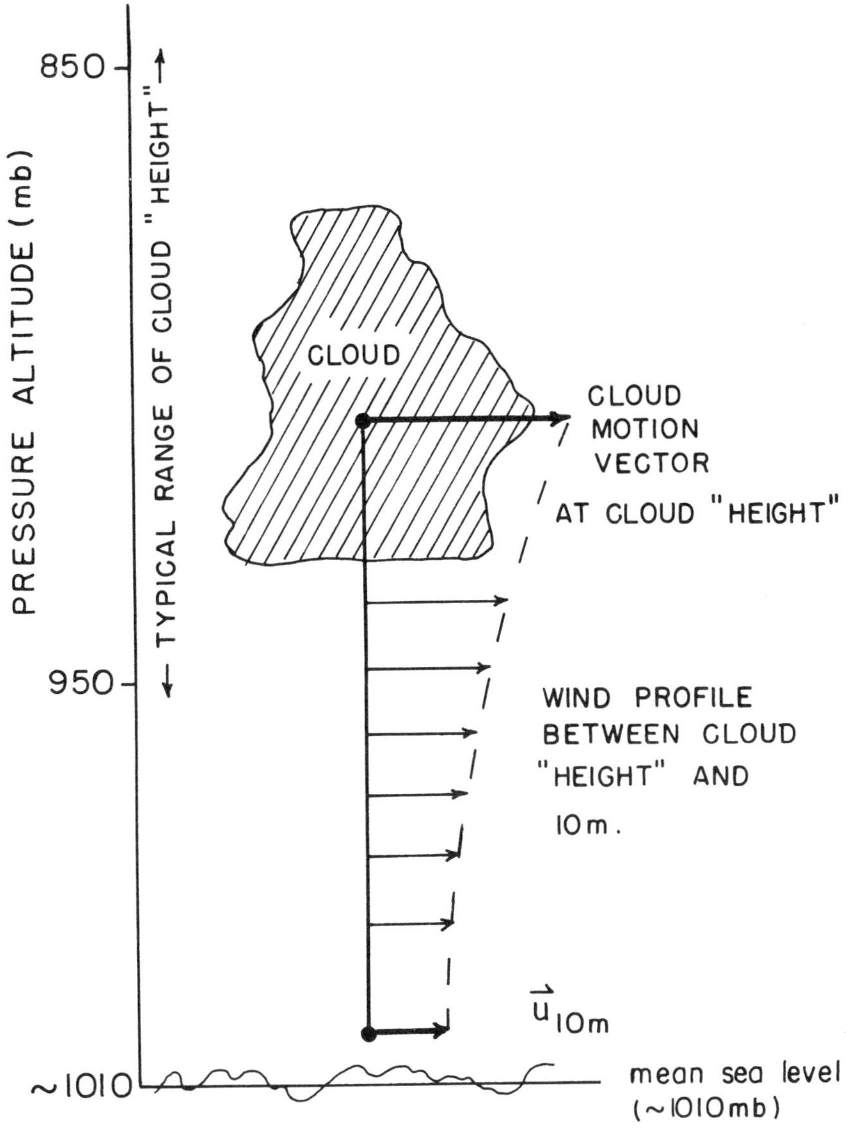

Figure 3. Schematic illustration of how 10m height winds are obtained from Cloud Motion Vectors (CMVs). CMV is obtained from some process and assigned to a fixed cloud height. This vector is adjusted to give a nominal 10m height wind vector by another empirical process. The actual height of a CMV and the shear profile of the wind corresponding to that CMV are not known. (See text).

cloud top and the surface can be large. On time scales of a day or so the shear in speed can be >10 ms^{-1} as well as very strongly variable in time, and up to 180 degrees in direction. According to comparisons between NOAA/NESS CMVs and buoy winds in the eastern tropical Pacific, there is very little relationship between the two winds on time scales shorter than 5 to 10 days but useful relations appear to exist for longer time scales (5). Even for monthly means, no single relationship appears likely to hold over large space scales. A single comparison of a tropical Pacific monthly mean wind field (Nov. 1979) derived independently (SSEC CMVs plus a fixed turning angle and speed reduction factor, Sadler-U. Hawaii NESS CMVs plus a climatological shear correction) shows RMS differences of ~2.5 ms^{-1} and maximum point differences of ~8 ms^{-1}, but generally excellent large scale spatial correlation was found (4).

C. Wind Stress Fields

For physical purposes a wind stress field, rather than a collection of observations in different locations and at different times, is desired. None of the data sources discussed above can provide basin scale coverage for many two-week average wind fields, although the problems are different for each data source. Buoy winds are far too few in number to give the needed spatial coverage, but give excellent sampling where they exist. Ship winds give coverage that varies in space and time and can be of poor quality and/or affected by aliasing; particularly in the tropics and the southern hemisphere where large areas are basically unsampled. CMV derived winds give good sampling and spatial coverage in many areas, but there is not basin-wide coverage or consistent data processing over the globe and the accuracy of adjusted-to-near-surface values is not well known.

Since over much of the globe good quality surface pressure information is available and can be a strong constraint on the surface wind, it seems natural to incorporate all of the above sources of data, plus pressure and try to produce a consistent global wind field. This is done on a routine basis in the analysis/forecast cycles at various national weather forecasting centers. These analysis/forecast cycle products do not have maximum accuracy of the surface wind field as a primary objective, but have to accommodate the demands of assimilating different types of data of variable quality into suitable "first-guess" fields for a numerical model forecast. Except under special circumstances (e.g., severe storms that cause shipping or coastal damage) the surface winds are not a primary test of forecast success. The accuracy of these wind fields is thus not well documented in the literature.

Three surface wind fields (Fleet Numerical Ocean Central pre-1981 "global band", Naval Environmental Prediction and Research Facility "NOGAPS" and National Meteorological Center "Optimum Interpolation") for the November 1979 average for the tropical Pacific have been compared (4). Although such a restricted comparison is clearly of limited utility, it was found that RMS differences between the fields were 1.0 to 2.5 ms^{-1}, with the two Navy fields tending to be more alike than either Navy and the NMC field. Point differences in these monthly mean fields could be as much as 6 ms^{-1}.

These same wind fields were also compared with the previously mentioned CMV-derived fields and with a ship wind-based field produced at Florida State University by J. O'Brien and the U. Hawaii CMV-derived field, modified to include some ship wind data. The RMS differences between the two U. Hawaii fields were the smallest (~0.5 ms^{-1}), but between the others values were ~2 ms^{-1}; differences between fields produced from different data sources or methods did not tend to be noticeably greater than between those produced from similar data or methods. Figure 4 shows the maps of zonal wind for Nov. 1979, from (4).

To put the magnitude of these differences into perspective is important. Monthly mean surface wind speeds in the tropics range from ~3 to ~8 ms^{-1}, and the seasonal changes in wind speed are typically 2 to 3 ms^{-1}. Interannual variations in wind speed can be as little as ~1 ms^{-1} to 5 or more ms^{-1}, depending on location. Relative to any of these values, RMS differences of 2 ms^{-1} and maximum point differences of 5 ms^{-1} are unsatisfactorily large, particularly since we have no wind field standard to establish whether any field is actually better than any other. Because the wind stress depends at least quadratically on wind speed, uncertainties of 2 ms^{-1} easily imply uncertainties of 50% in wind stress in many locations.

These remarks may be unduly pessimistic for much of the northern hemisphere middle latitudes, because far more ship data is available there and surface pressure there much better constrains the winds than it does in the tropics. Comparisons for the middle latitudes in both hemispheres are now being carried out; but the lack of widespread reliable surface truth information will also limit the inferences possible from these comparisons.

D. Scientific Requirements

No short review can hope to provide more than a rough indication of the degree to which stress needs to be known to

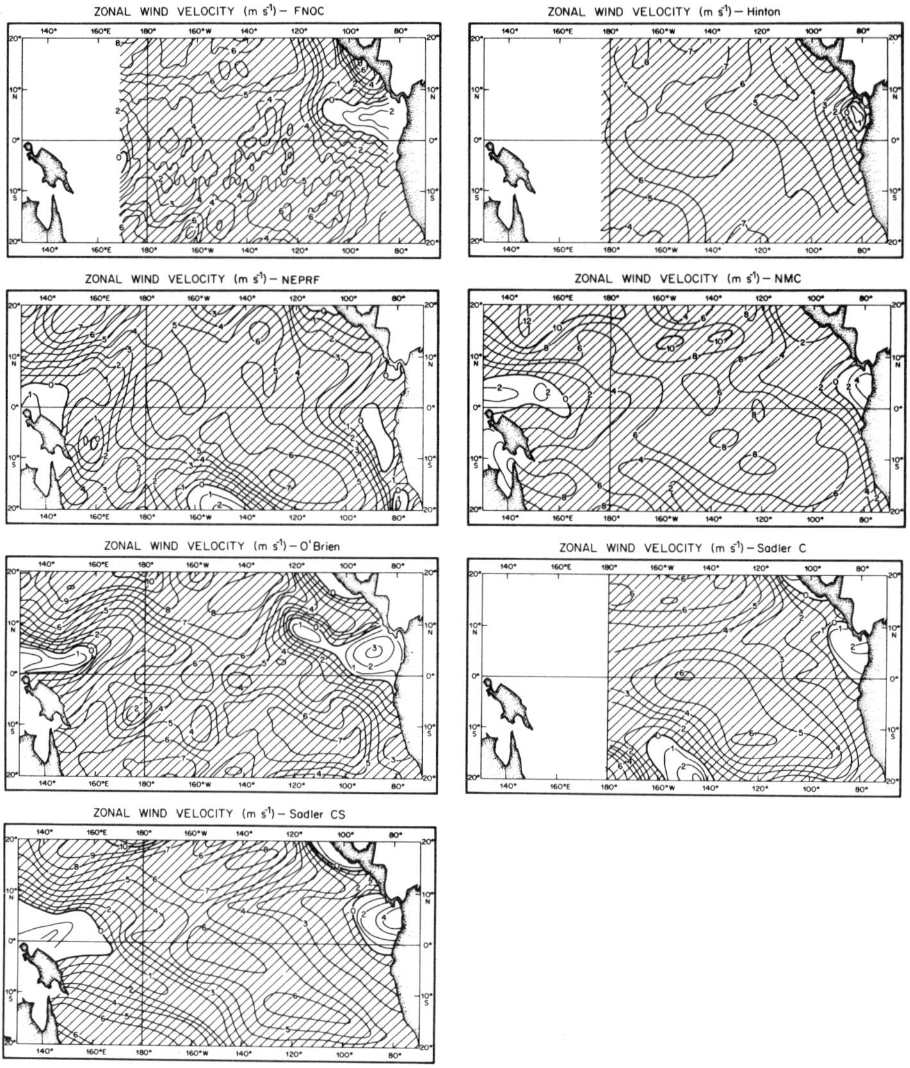

Figure 4. Maps of average zonal wind for the tropical Pacific for November 1979, from seven independent procedures (4). FNOC-U.S. Navy "Global Band" analysis/forecast; Hinton-SSEC CMV; NEPRF-U.S. Navy "NOGAPS" analysis/forecast; NMC-U.S. National Meteorological Center "O.I." analysis/forecast; O'Brien-ship wind analysis; Sadler C - NOAA/NESS CMVs adjusted with climatological wind shear; Sadler CS - as Sadler C, but also incorporating ship winds.

meet a few scientific objectives. In fact, because stress is poorly known at present, future information may reveal that any such estimates were deficient. Still, it seems useful to examine briefly the issue here, as it relates to some problems of the low frequency ocean circulation.

In the tropics monthly mean zonal wind stresses generally range from nearly zero to about 0.1 Nm^{-2}, with annual changes ranging from ~0.01 to ~0.1 Nm^{-2}. Interannual changes can be larger than climatological annual changes in the equatorial Pacific; elsewhere they appear smaller. If it is desired to understand both the normal seasonal cycle and the interannual events, the stress must be known well enough to resolve the changes in both situations. Were the monthly mean stresses known to 0.01 Nm^{-2} with resolution of ~1° in latitude and 5° in longitude much could undoubtedly be learned, but knowledge to 0.001 Nm^{-2} on 1° by 1° resolution would be far preferable. At the latter level of accuracy, predictions from theoretical and numerical models of ocean response could no longer rationalize poor forcing data as the source of their discrepancies with data.

The more stringent limit is important for another reason. Wind stress curl, as well as wind stress, is an important part of the forcing of the ocean circulation. In the tropics the curl extrema are associated with the main convergence zones, and estimates of the maximum curl are ~10^{-7} Nm^{-3}, with maximum annual changes of roughly the same amount. Elsewhere the maximum and annual changes are smaller by an order of magnitude. Stresses known to 0.001 Nm^{-2} and with 1° resolution provide curl values adequate to allow study of their important effects.

Recent studies with Pacific atol wind data suggest that time resolution better than that allowed by monthly mean values may be important for understanding the initiation of El Nino events (6). Five to ten day average data with accuracy of ~0.03 Nm^{-2} on a 2° x 2° grid may be adequate for initial studies of the effects of the westerly wind bursts they describe. Sea level records clearly show oceanic response associated with wind events on these time scales, but their larger scale roles remain unclear. These shorter time scales must also be resolved if the oceanic effects of monsoon outbreaks, or the onsets of the Indian Ocean monsoon are to be studied.

The low frequency response of the middle latitude ocean circulation appears to be more strongly forced by wind stress curl than by stress. The maximum monthly mean stress is ~0.2 Nm^{-2}, with maximum annual changes of ~0.1 Nm^{-2} in the North Atlantic. The curl maximum appears to be ~4 x 10^{-7} Nm^{-3} with annual changes of ~10^{-7} Nm^{-3}; over much of the basin the curl and its annual changes are ~10^{-8} Nm^{-3} (7). To resolve curl and curl changes of the smaller values calls for stresses to be accurate to roughly 10^{-4} Nm^{-2}, if the grid is to be 1° x 1° in space.

Many other sorts of accuracy constraints could be discussed. Although not of direct oceanographic concern, better knowledge of the divergence of the surface wind is very important for air-sea interaction studies in the tropics. It may be an essential key to unlocking the secrets of the El Nino-Southern Oscillation cycle, because convergence is so strongly coupled to heating, and heating appears to be such a major factor in the atmospheric forcing of parts of the ENSO cycle. Maximum monthly mean surface convergence in the tropics appears to be a few times 10^{-6} s^{-1} and the big changes during the ENSO cycle can be of this order.

This last requirement offers a very different sort of constraint than do the earlier ones. Any vector field can be decomposed into the sum of a field with zero curl (the divergent part) and of one with zero divergence (the rotational part). For a variety of reasons, analysis/forecast cycle procedures often find it practical to minimize the divergent part of the flow field in their analyses. Cloud motion vectors, on the other hand, appear well suited to identify convergent flow. The need to correctly identify both parts of the wind is clear.

The above requirements will only be attainable if much of the error in the individual observations is random. If there are systematic errors to any substantial extent, these requirements cannot be met. Randomness in space, as well as in time, can be very important. For a variety of physical problems in tropical oceanography, it is the product of the error in stress and the distance over which the error persists (the decorrelation length of the stress error) that enters error estimates for the oceanic response, not just the stress errors (8). This is another reason for striving to get independent data with high spatial resolution. At present it is very difficult to estimate the length scales of errors in the wind stress. The results of (4) suggest that the tropical scales for monthly means may be tens of degrees of longitude.

No mention has been made of the constraints imposed by surface wave or current forecasting, weather forecasting, *etc*. These areas require information outside the time scales under consideration here, and have very special needs that should be given detailed discussion elsewhere.

E. The Future

From the above it is clear that our knowledge of the surface wind stress is deficient in many regions and on many space and time scales. Fortunately a variety of new or recently developed technologies appear likely to be able to greatly improve our knowledge, if their promise is borne out by further testing and if they are appropriately deployed. In this short space it is only possible to touch briefly on a few issues important for progress.

The first area concerns better knowledge of the relationship between stress and near surface wind. Better capability to measure stress from shipborne instruments or buoys would greatly assist in refining and extending the bulk formulae as well as documenting their deficiencies under special conditions. The recent success in reconciling dissipation method and Reynolds method stress measurements encourages the belief that stress measurements are both improving and becoming more straightforward to make (1). It is essential to keep in mind that stress is the physically important quantity, not wind. Winds themselves are largely oceanographically unimportant except insofar as they can be turned into stresses (although see Section D). There is no substitute for more and better stress data.

Until accurate direct measurements of stress become easily made, it will be necessary to work with wind and boundary layer stability data to estimate stress through a bulk formula. At present it appears likely that our knowledge of the wind and of the air-sea temperature difference is poor enough over most of the world ocean that these uncertainties not uncertainty in the bulk formula, sets the main limits on present knowledge of wind stress. It is thus also important to improve our knowledge of the winds and air-sea temperature differences as much as possible. Better knowledge of these quantities is possible in many ways.

Many of the ships and islands that record surface marine observations do not report their data to the world weather forecasting network or even to national climate center archives. Improved collection of these data is possible and must be undertaken; (9) shows that it was possible to double the amount of ship data available to discuss the 1972 El Nino period through determined data collection efforts and (10) also finds that large amounts of collected data remain outside the data centers. Even when data are collected at national climate archives, funds may not be available to process the data into computer-useable form on a timely basis, or national policy may discourage dissemination of the data to the global research community.

Unfortunately, over much of the world ocean there is limited ship traffic and it could not provide enough data even if all of the data were accurate and made available. The FGGE experience suggests that technologically straightforward pressure telemetering drifting buoys can significantly change (in some areas) the winds produced via analysis/forecast center methods, and such buoys appear to be the least costly way to improve southern hemisphere surface wind information. If satisfactorily accurate wind or stress sensors are developed for use on drifting buoys, such buoys could also greatly improve our knowledge in the tropics as well as in the southern hemisphere. But the density of observations will likely remain small enough that analysis/forecast type data assimilation often will be needed to produce fields of wind or stress.

The need is great for sensors that can provide data with global coverage and of acceptable accuracy. Only satellite borne sensors can give the desired coverage; fortunately, microwave sensors show promise of being able to give acceptable accuracy. The type of instrument which has received the most attention lately is a multi-antennae microwave radar, usually called a scatterometer. K-band or C-band microwave radiation is strongly scattered by the centimeter scale structure of the sea surface, and the surface scattered signal can be interpreted to obtain 10m height neutral wind speed and wind directions(s) (11).

It is important to note that, in general, a unique direction is not obtained from the inversion process. Depending upon the configuration of the instrument and the surface characteristics being sampled, there can be several directional "ambiguities" in the solution. Selecting the correct direction out of the set of possible directions (which has generally been called the "de-aliasing" problem) has proven to be non-trivial from the SEASAT scatterometer data set. Recent design studies indicate that the ambiguity problem can be greatly reduced by altering the number and orientation of antennae on future instruments (11), (12).

Past scatterometer design specifications have called for determination of wind speed and direction. The relationship between the received radar backscatter and wind speed is the subject of much current research. At present the relationship is empirically determined, and of rather simple form. See (11), (12) for an introduction to SEASAT-style scatterometry and references. There remain questions about the form of the best empirical relationship, and the theoretical underpinnings of the relationship largely remain to be established. Also unclear is the degree to which backscatter is affected by the ambient surface wavefield and other factors.

For maximum oceanographic utility it is more desireable to obtain stress than wind data from the processing of radar backscatter. Because of the paucity of ocean surface stress measurements, it is more difficult to develop the needed empirical relationship. However, with the recent improvement in ability to measure stress well, it may now be feasible to plan for the development of a backscatter-stress relationship.

The sampling constraints discussed in section B will continue to impose constraints on how well stress is known, even if the scatterometer concept is developed for stress and a satellite mission is undertaken. The likely temporal repeat coverage of a single instrument does not appear adequate to avoid all problems of high frequency aliasing of lower frequencies (11). Again, some sort of data assimilation process may be needed to produce the most scientifically useful data set. Only with multiple satellites would sampling be frequent enough to produce minimal aliasing on a global basis. Spatial resolution appears to be largely a question of antennae size, available power and accuracy requirements - 10 to 25 km spatial resolution appears feasible for the next generation scatterometer instruments. This is far superior to the resolution available from any other source. Appropriate amounts of high quality data from scatterometer(s) could revolutionize our knowledge of the surface wind and stress fields.

Other satellite-borne sensors working at radar frequencies appear able to give wind and/or stress information to varying degrees. Analysis and development work on these sensors is continuing, but some sort of scatterometer appears most likely to be the first instrument to give useful global coverage.

There is an urgent need in physical oceanography, meteorology, marine operations and climate research for improved surface wind and wind stress information. Much can be done to improve our knowledge through better collection of existing observations, deployment of properly instrumented drifting buoys, better calibration and collection of cloud motion vectors, increased scrutiny and development of analysis/forecast methods of surface data assimilation and more widespread direct measurement of surface stress. But the space and time sampling constraints on such techniques will inevitably limit the improvement to less than is desireable or needed. Only satellite-borne sensors, adequately deployed and calibrated against high quality surface data can provide the needed information. The microwave scatterometer shows promise for meeting the required accuracy needs, and may soon revolutionize our knowledge of the surface wind and stress fields.

F. Acknowledgements

This work was supported by NASA grant NAG 5-322 and NSF grants OCE-83-01787 and OCE-79-21815. Many individuals have freely offered their perspectives on the issues discussed here; my thanks to all, and especially to W. Large, J. Sadler, W. Pierson and D. Chelton.

REFERENCES

(1) Large, W.G. and S. Pond, 1981, Open ocean momentum flux measurements in moderate to strong winds, J. Physical Oceanog., 11, 324-336.

(2) Smith, S.D., 1981, Coefficients for sea-surface wind stress and heat exchange, BI-R-81-19, Bedford Inst. Oceanog., Dartmouth, Nova Scotia, Canada.

(3) Willebrand, J., 1978, Temporal and spatial scales of the wind field over the North Atlantic and North Pacific, J. Phys. Oceanogr., 8, 1080-1094.

(4) Halpern, D., D.E. Harrison, et al., 1981, Intercomparison of tropical Paciofic mean November 1979 surface wind fields, Dept. Meteorology and Phys. Oceanog., Report 82-1, M.I.T., Cambridge, MA 02139.

(5) Halpern, D. and R. Knox, 1983, Coherence between low level cloud motion vectors and surface wind measurements near 0°N, 152°W from April 1, 1979 to Feb. 1980 Atmos.-Ocean, 21, 82-93.

(6) Luther, D., D.E. Harrison and R. Knox, 1983, Zonal winds in the central equatorial Pacific and the onset of El Nino, Science, 222,

(7) Hellerman, S. and M. Rosenstein, 1983, Normal monthly windstress over the world ocean with error estimates, J. Phys. Oceanogr., 13, 1093-1104.

(8) SEQUAL Wind Document, 1981, SEQUAL Office, Lamont-Doherty Geological Observatory, Palisades, NY.

(9) Ramage, C.S. et al., 1980, Meteorological Atlas of the 1972-73 El Nino, Report UH Met 80-03, Department of Meteorology, U. Hawaii, Honolulu, HI.

(10) Cutchin, D.L., 1983, Numbers of meteorological observations by ships in the tropical Pacific, Report SIO 83-9, UC-SD, La Jolla, CA.

(11) Satellite Surface Stress Working Group, 1982, Scientific opportunities using satellite wind stress measurements over the ocean, NASA HQ, Oceanic Processes, Washington, D.C.

(12) Pierson, W., Jr., 1983, Highlights of the SEASAT-SASS program: A review, Ch. 4, in Satellite Microwave Remote Sensing, ed. Allan, Ellis Howard Ltd., Chichester, Eng., April 1983.

WIND SPEED AND STRESS OVER THE OCEAN: SCATTEROMETER VERSUS SURFACE MEASUREMENTS

W. G. Large

National Center for Atmospheric Research
Boulder, Colorado, USA

ABSTRACT

Once routine, global wind measurements of known accuracy become available from satellites, both oceanography and meteorology will be relieved of a great observational burden. The SEASAT experience demonstrated that its scatterometer system, SASS, could do the job, but the following problems still exist:

i) The derivation of wind speed or stress requires a purely empirical algorithm.

ii) There is evidence of systematic biases in the SASS winds.

iii) The SASS algorithm is based on mid-latitude surface measurements during summer.

iv) Differences in SASS and surface winds are sometimes attributed to spatial versus time averaging.

v) It seems likely that backscatter depends on more parameters than have been considered in retrieval algorithms.

It is argued that at this stage, the next likely scatterometers, on NROSS (KU-band similar to SEASAT) and certainly ERS-1 (C-band), will both require surface measurements on which to base or improve their retrieval algorithms. Since the ocean surface is still a "black box" as far as the backscatter relationship to the wind is concerned, it is further argued that one algorithm is required to derive wind speed and another to derive wind stress. A measurement program is then outlined, which would improve upon the present state of scatterometer anemometry, by providing spatially averaged wind speed and stress at a variety of oceanic sites, together with the necessary supplementary data.

C. Gautier and M. Fieux (eds.), Large-Scale Oceanographic Experiments and Satellites, 117–124.

1. INTRODUCTION

Routine satellite observations of the over ocean wind field (velocity or stress) is an exciting near future prospect, the concept of which was proven by the SEASAT scatterometer, SASS [1,2]. This possibility is of added significance, since wind is the single most important parameter needed for both latent and sensible heat flux estimates. With present technology the wind forcing could be observed up to global spatial and climate temporal scales. Conventional surface measurements are neither as accurate nor as large scale. Wind reports (visual estimates) from the Voluntary Observing Fleet (VOF) do cover the northern mid-latitude oceans well, but they are not global [3], and have unknown errors that are likely to be systematically more than 10% in speed [4] (~30% in stress). Accuracies of 3% in speed and 10 to 15% in stress may be obtainable from research ships and buoys [5], but because of the costs, such measurements probably could not be extended beyond a few hundred kilometers and about one year.

In view of the above considerations, it behooves us to strive towards obtaining the best possible satellite scatterometer winds. Certainly existing instrument designs and algorithms are capable of greatly improving our present wind observations [6], but they may not be adequate for certain important studies and programs. As a specific example, in order for monthly or longer mean wind stresses to be considered as "known" inputs to ocean climate models, they must be measured to within 20% [7]. In addition the random error needs to be minimized and understood, so that the scatterometer winds can be used on as small a spatial scale and as short a temporal scale as possible, thus relieving many experimental programs of difficult and costly surface measurements, and providing models with the broadest band forcing function possible.

In order for scatterometer winds to approach their potential accuracy, resolution, and hence scientific value, it would seem to be essential to discover the important parameters that govern backscatter and to calibrate *in situ*, at least the next few spacecraft systems. In this regard, accurate and reliable surface measurements of wind velocity and stress as well as potentially important parameters such as sea surface temperature, surface waves, surface contaminants and the local meteorology are needed.

2. SCATTEROMETER STATUS

Microwave scatterometry is a rapidly developing field due mainly to the numerous evaluations of SASS and airborne scatterometers and to the designing of the next spacecraft systems. A brief summary of the important issues to emerge follows, but it may not reflect the results of ongoing studies.

i) Present retrieval algorithms are purely empirical, and as such have been subjected to a generally unknown degree of tuning [1,8]. The scatterometer, there fore, remains a secondary wind sensor, whose uncertainty must be at least as large as that of the calibration data used as the standard.

ii) There are indications that SASS wind speeds may be biased high (>6% vs FNOC winds [9]), (~10% vs Tropical Pacific ship winds [10], though there is a great deal of uncertainty in the comparative data.

iii) The SASS algorithm has now been tuned to JASIN wind velocity data [8], which represents only one geographical area (60°N, 12°W) and one season (summer, 1978). The uncertainty in the JASIN wind speeds, although much less than in VOF estimates may be greater than could potentially be achieved from scatterometer winds [5].

iv) Conventional wind observations and satellite derived winds contain different components of the mesoscale variability and hence should not be identical. The data processing of each technique should be such as to make them as compatible as possible before intercomparison or calibration [11].

v) The return of K-band scatterometer radiation from the ocean is governed by resonant scatter from short waves tilted by long waves. The backscatter coefficient, σ, has been found to depend both on wind speed and direction [12], hence the success of wind velocity retrieval algorithms. However, listed below are several other parameters that could be important and their effects on σ need to be understood if systematic biases and erroneous winds are to be avoided.

a - wind stress or the friction velocity, u* [2], as is further discussed in Section 3.

b - near surface atmospheric stability, ζ, as discussed by [13].

c - sea surface temperature, Ts, through its influence on stability and, probably more importantly, its effects on viscosity and surface tension and hence on the spectral density of short waves, [13].

d - surface contaminants (organic), that could be concentrated at surface convergences.

e - the non-local wave field (swell), by altering the short wave statistics.

f - rainfall, either by roughening the sea surface or by scattering while falling. Resulting anomalous SASS winds in a region of deep mid-level convection are discussed by [14].

vi) Scatterometers are stable with demonstrated good absolute calibration and measurement precision [1], so that instrumental errors can be made negligible (by use of the Amazon rain forest as a standard in the case of satellite systems).

3. EMPIRICAL ALGORITHMS

For a given surface stress $\vec{\tau}$ of magnitude $\rho u*^2$ (ρ is air density), the mean wind speed profile for heights, $z \gg z_o$ is

$$U(z,\zeta) = u*\kappa^{-1} (\ln(z/z_o) + \psi(\zeta)), \qquad (1)$$

where von Karman's constant, κ, and the flux profile function $\psi(\zeta)$ are empirically determined and the surface roughness, z_o, is usually , but not always, uniquely related to U at ζ = 0 through the neutral drag coefficient $C_D{}^N$ (z) = $\kappa/\ln(z/z_o)$ = $(u*/U(z,0))^2$. The SASS model function relates σ(db) to wind speed (m/s) and direction:

$$\sigma = G(\theta,\chi,\varepsilon) + H(\theta,\chi,\varepsilon) \log_{10} U(19.5,0) \qquad (2)$$

where G and H are empirical functions of incidence angle, θ, the aspect angle, χ (the antenna azmith-wind direction) and the polarization (either horizontal or vertical). If (2) is truely a complete description then it appears as if the random error in U(19.5,0) is less than ±2m/s [1]. However, the shift to a real wind $U(z,\zeta)$, when ζ is unknown leads to additional errors of up to 10% for $-1.0 < \zeta < .2$ [15]. If a wind stress is to be calculated then a form of $C_D{}^N$ (19.5) must be assumed, which has a 10% uncertainty and there is an additional error of up to 10% at winds above 10 m/s where $C_D{}^N$ depends on U, so that the total error in the stress magnitude from 10 to 20 m/s is nearly 40%. This error is not entirely random and does not include errors due to parameters not contained in (2) or due to bias in the U(19.5,0) values used to determine G and H in (2). Such a bias could arise both from inaccurate $U(z,\zeta)$ measurements and in shifting them to 19.5 m and neutral stability.

The friction velocity near the surface is nearly constant and a more appropriate model function may be

$$\sigma = G^*(\theta,\chi,\varepsilon) + H^*(\theta,\chi,\varepsilon) \log_{10} u^* . \quad (3)$$

Of course, (2) and (3) will be directly related if u^* and U(19.5,0) are uniquely related by C_D^N (19.5). It is not possible to measure over ocean values of u^* (10%, [15]) as accurately as surface winds $U(z,\zeta)$ (3%, [5]), but because of the need to shift to U(19.5,0) and the problems associated with C_D^N (19.5), there are advantages to using (3). A very limited data set suggests that σ correlates equally well with U(19.5,0) and u^* and that (2) is preferable if wind speeds are required and (3) is preferable if wind stress is wanted [2]. There do appear to be situations associated with non-equilibrium wind/wave conditions when the U(z,0) to u^* relationship breaks down [15,16]. In order to deal with such cases it is necessary to understand whether backscatter depends more on wind speed or wind stress (u^*) or if both are involved.

4. EXPERIMENTAL PROGRAMS

This section outlines two surface measurement programs designed to advance scatterometry by addressing the issues raised in the previous sections. At present a NASA airborne scatterometer exists and there are plans for spacecraft systems on ERS-1 (C-band), NROSS (KU-band) and Japanese satellites. Firstly, it seems prudent, in view of the order $100 million cost of each system, to initiate pre-launch investigations into the more outstanding scatterometer issues with the airborne system (KU and C-bands). Secondly, post launch verification and calibration experiments are needed, because these satellite systems will not be identical to SASS, and there is evidence that the SASS calibrations are not universally applicable (see also the concluding remarks in [1]).

The Jet Propulsion Laboratory, JPL, has proposed to conduct aircraft scatterometry in conjunction with boundary-layer instrumented aircraft flights [17]. Independently, tentative plans are being made for an air-sea interaction experiment near Bermuda [18], which has many of the surface and boundary-layer components required for the JPL program. Negotiations are underway to coordinate the two experiments, so that there would be <u>all</u> the potentially important surface measurements, sufficient parameter variation (e.g., SST range), a degree of redundancy and a number of well planned scatterometer and boundary-layer aircraft flights. The relevant components might be the following:

i) A six-month array of five surface moorings, each with two proven meteorological packages. The mooring wind speeds may approach an accuracy of 3% and they can be cross-calibrated so that very good spatially averaged winds are measured over a scatterometer footprint.

ii) Shipboard dissipation measurements [15] of the wind stress, velocity and a "local" drag coefficient, which, when applied to the moored array wind speeds, can give a spatially averaged wind stress.

iii) Aircraft turbulent stress measurements in the boundary layer, while surveying wind velocity, SST, air temperature and humidity horizontally.

iv) Ship launched radiosondes sampling in the vertical and giving the atmospheric liquid water content.

v) Shipboard boundary-layer turbulence measurements using SODAR (Sound Detection and Ranging) and LIDAR (Light Detection and Ranging), which would increase confi dence in the spatial averages of wind and stress and especially in the aircraft stresses.

vi) Sea surface temperature measurements (20° to 29°C) from the moorings, a ship, an aircraft and AVHRR infrared satellite imagery.

vii) Surface wave measurements; long wave (1-500 m) with directional information (Pitch-and-Roll Wavebuoy, LASER Profilometer) and short wave (.01-1 m) (Buoy mounted resistance wavestaff) [17].

viii) Visual observations of surface contamination.

With such a suite of observations it would be possible to answer, in large part, the question; What parameters determine backscatter and which do not?

The strategy for post-launch surface observations would depend a great deal on the results of a pre-launch airborne scatterometer study, which, therefore, ought to plan for an accelerated data analysis phase. At the moment a possibility would be to have a dedicated ship tending an array of wind measuring platforms (moored or drifting). The ship would deploy the array over a scale corresponding to a satellite scatterometer footprint, cross-calibrate the wind sensors, and directly measure the surface stress and a local drag coefficient. Throughout the lifetime of the satellite systems (18-36 months) the array would have to be moved to all the potentially different backscattering regions and seasons as suggested by the airborne scatterometer studies. Certainly more than one such program could be run if it is necessary to increase geographical coverage. Given the present focus on the tropics, because of its seemingly large impact on global climate, this region would probably receive a high priority, but other remote areas such as the Southern Ocean could also be important and should not be ignored.

A concluding remark is that global scatterometer winds also could be used to enhance our knowledge of past surface winds. Sea level pressure observations and daily analyses have been relatively extensive, since 1899 in the northern hemisphere and for 25 of the past 33 years in the southern. If techniques are developed to estimate wind stres from these data and verified by comparison with global scatterometer winds of known accuracy (as they have been using North Atlantic weathership winds [19]), then we could have considerable confidence in past winds using the same techniques.

REFERENCES

(1) Jones, W. L., *et al.*, 1982: The SEASAT-A Satellite Scatterometer: The Geophysical evaluation of remotely sensed wind vectors over the ocean. *J. Geophys. Res.*, *87*, C5, pp. 3297-3317.

(2) Liu, W. T., and W. G. Large, 1981: Determination of surface stress by SEASAT - SASS: A case study with JASIN data. *J. Phy Oceanogr.*, *11*, pp. 1603-1611.

(3) Taylor, P. K., 1983: The measurement of the ocean surface fluxes of sensible heat, latent heat and water for WOCE. World Ocean Circulation Experiment Report.

(4) Graham, A. E., 1982: Winds estimated by the Voluntary Observing Fleet compared with instrumental measurements at fixed positions. Met. Mag., *111*, pp. 312-327.

(5) Weller, R. A., R. E. Payne, W. G. Large and W. Zenk, 1983: Wind measurements from an array of oceanographic moorings and from F. S. Meteor during JASIN 1978. Submitted to *J. Geophys. Res.*

(6) Scientific opportunities using satellite surface wind stress measurements over the ocean. Report of the Satellite Surface Stress Workshop Group, J. J. O'Brien, Chairman, for NASA, June 1982.

(7) F. P. Bretherton, personal communication, 1983.
(8) Seasat Data Utilization Project Report. NASA, Jet Propulsion Laboratory, Pasadena, No. 622-235, July 1982.
(9) D. B. Chelton, personal communication, 1983.
(10) Chelton, D. B., J. J. O'Brien, 1982: Satellite microwave measurements of surface wind speed in the Tropical Pacific. Tropical Ocean-Atmosphere Newsletter, 11.
(11) Pierson, W. J., Jr., The measurement of the synoptic scale over the ocean. J. Geophys. Res., 88, C3, pp. 1683-1708.
(12) Moore, R. K., and A. K. Fung, 1979: Radar determination of winds at sea. Proc. IEEE, 67, pp. 1504-1521.
(13) Liu, W. T., 1983: The effects of the variations in sea surface temperature and atmospheric stability in the estimation of average wind speed by Seasat - SASS. Submitted to J. Phys. Oceanogr.
(14) Guymer, T. H., J. A. Businger, W. L. Jones, and R. H. Stewart, 1981: Anomalous wind estimates from the Seasat scatterometer, Nature, Lond. 294, pp. 737-739.
(15) Large, W. G. and S. Pond, 1981: Open ocean momentum flux measurements in moderate to strong winds. J. Phys. Oceanogr., 11, pp. 324-336.
(16) Byrne, H. M., 1982: The variation of the drag coefficient in the marine surface layer due to temporal and spatial variations of the surface wind and sea state. Ph.D. Thesis, University of Washington.
(17) Freilich, M., JPL Scatterometer airborne experiment program presented by JPL, June 1983.
(18) R. A. Weller, personal communication, 1983.
(19) Thompson, K. R., R. F. Marsden and D. G. Wright, 1983: Estimation of low-frequency wind stress fluctuations over the open ocean. J. Phys. Oceanogr., 13, pp. 1003-1011.

A SUMMARY OF THE WIND DATA AVAILABLE FROM SATELLITES FROM THE PAST HISTORY TO FUTURE SENSORS

Donald Wylie and Barry Hinton

Space Science and Engineering Center
University of Wisconsin-Madison
Madison, WI 53706, U.S.A.

ABSTRACT

The currently available satellite methods for obtaining wind information over oceans are discussed with respect to the time and spatial coverages of each method and its accuracy. At the present, cloud motion data is being taken for most of the world's tropical and mid-latitude oceans. This record extends back to 1975 for the western hemisphere. Full coverage of all oceans started in 1979. Wind information can be obtained from passive microwave sensors (thermally emitted microwaves) for all oceans from 1978 to the present. Radar altimeters and scatterometers also provide wind data.

The future outlook is for continued cloud motion coverage. The current passive microwave sensor may terminate before its successor is launched in 1985. A radar altimeter is scheduled for launch in 1984 and two scatterometers for 1989.

Only two methods provide wind direction information, the cloud motions and scatterometers. To correct for this deficiency, along with other deficiencies in coverage, a combined analysis using all available data is suggested.

1. INTRODUCTION

There are five different types of satellite data from which wind information can be extracted over oceans. These include visible and infrared imagery from operational meteorological satellites and microwave and radar sensors flown on experimental satellites. This paper will discuss the relative advantages and disadvantages of each wind sensing method and outline what the prospects are for wind information on future satellites.

C. Gautier and M. Fieux (eds.), Large-Scale Oceanographic Experiments and Satellites, 125–146.

The five methods are: 1) cloud motion measurements, 2) the sun glitter reflection from the sea surface, 3) measurements of thermally emitted microwaves from the ocean surface, 4) the reflection of radar pulses off the surface from an altimeter, and 5) radar reflections from a multi-directional scatterometer. These methods are summarized in Table 1 and are described in order below. The relative characteristics of each method are described in Section 7 and Table 2 and the accuracies in Section 8 and Table 3. The length of record available and future prospects are summarized in Section 9.

2. CLOUD MOTIONS

Cloud motions have been widely used for obtaining wind information because they can be obtained from the operational satellites. A running climatology of surface winds over the tropical Pacific Ocean has been compiled at the University of Hawaii by (1). Clouds are routinely tracked on the Geostationary Operational Environmental Satellites of the United States (GOES East and West) by the National Oceanic and Atmospheric Administration's National Environmental Satellite and Data Information Service (NOAA/NESDIS) for the area from 30°W to 180°W longitude. Three wind sets/day are obtained.

Cloud motions also are routinely tracked from the geostationary Meteosat images by the European Space Agency (ESA) in the area from 50°W to 50°E longitude and also from the Japanese Geostationary Meteorological Satellite (GMS), from 80°E to 180°E longitude operated by the Japanese Meteorological Agency (JMA). These four satellites cover 80% of the tropical oceans.

The Indian Ocean was covered by a United States geostationary satellite during 1979 for the First GARP Global Experiment (FGGE) for a one-year period. Coverage will resume in late 1983 with the launch of Insat by the Indian Government.

The obvious problem with using cloud motions for estimating surface winds is the vertical wind shear in the atmospheric boundary layer. Fortunately, tropical cumuli move with the wind at their bases and not their tops. This has been extensively documented with simultaneous measurements of wind by aircraft and satellite cloud motions (2). The sub-cloud shear has been

Table 1: Methods of Monitoring Winds from Satellites

Method	Type of Satellite: Geo-Stationary	Type of Satellite: Polar	Type of Instrument Used	Wind Direction
Cloud Motions	Yes	No	Visible & IR Images	Yes
Sun Glitter Reflections	Yes	Yes	Visible Images	Maybe
Thermal Microwave	No	Yes	Multi-channel Microwave Radiometer (3-37 GHz)	No
Radar Backscatter	No	Yes	Altimeter	No
Radar Backscatter	No	Yes	Bi-directional Scatterometer	Yes

removed using mean statistics from comparisons of the low level cloud motions (below 700 mb) to surface ship wind reports. The Hawaii method uses mean difference between monthly climatologies of cloud motions and ship reports for deriving a statistical wind shear which is geographically variant for each month of the calendar. For FGGE, the University of Wisconsin has developed similar cloud-ship wind shear statistics (3, 4). However, the Wisconsin wind shear statistics show direction and speed categories as well as geographical variances in an attempt to isolate weather situation conditions that affect the shear.

Table 2: Sampling Limits of Satellite Wind Monitoring Systems

Method	Resolution Available Time	Space	Sampling Restrictions
Cloud Motions	Minutes *2/day	100 km *500 km	1. Not under High Clouds 2. Not in Clear Areas
Sun Glitter	1/day	10 km	1. Limited to Latitudinal Belt under Sun 2. Clear or Broken Cloud cover
Thermal Microwaves	1/day	100 km	1. Satellite's Orbit-to-Orbit Gaps 2. No Heavy Rain
Radar Altimeter	1/day	**14 km	1. Orbit Samples Only Directly Under Spacecraft 2. No Heavy Rain
Directional Scatterometer	1/day	10-50 km	1. 1200 km Wide Swath Orbit-to-Orbit Gaps 2. No Heavy Rain

* Operational data products
** In the along-track direction

The end products of both Hawaii and Wisconsin are composite wind analyses that include both ship and cloud data. Graphical or statistical methods are used in the data combinations. By compositing the data, these methods produce wind analyses that are, for most purposes, better than the conventional non-satellite data because the imperfect coverage of each data source is often complementary (4).

3. SUN GLITTER REFLECTIONS

The strength of the sun's reflection off the ocean surface and the width of the reflection are modulated by the roughness of the surface which is produced by the wind. These factors were statistically quantified (5) using photographs from aircraft even before weather satellites were flown. Methods for using satellite images of the sun's "glitter" were proposed for both geostationary satellites (6) and polar orbiting satellites (7), but have not been routinely used at this time.

These methods require images of the ocean surface in relatively cloud free areas where the sun glitter reflection can be seen. From satellite altitude, however, the winds can vary over the large area that the glitter covers. Winds can be estimated at points in the image from any of three methods: 1) the magnitude of the reflection, 2) the change in brightness of a point in time as the sun passes, or 3) the gradient in brightness over a small area where constant winds can be assumed. The first method requires an accurate measurement of the image brightness. The sensor must be calibrated and the field of view must be free of obscurations from clouds, haze, and aerosols to accurately measure the intensity of the reflected sunlight. Because of these stringent requirements, most experiments have used the shape of the glitter pattern rather than its intensity at a single point.

The width of the glitter reflection is very narrow for calm or flat sea conditions, acting like a mirror, and broadens as the sea roughens in higher winds. A time sequence of geostationary images can be used to measure the width of the glitter's reflection as it passes a point, while with polar orbiting satellites, only one image/day can be obtained so the glitter width information has to be extracted from the horizontal brightness gradient over a small area where constant winds are assumed.

Wind direction measurements are theoretically possible from the elongation of the sun glitter reflection. Satellite techniques have not shown any skill in resolving the elongation because of the variations of the wind over the large area covered by the glitter pattern and the methods in which they are forced to sample, as previously discussed.

4. THERMALLY EMITTED MICROWAVES

In recent years, sensors for measuring the thermally emitted radiation in the microwave portion of the spectrum have been flown on experimental satellites. This radiation is mostly modulated by the temperature and emissivity of the earth's surface. The emissivity over oceans is strongly affected by the surface roughness and the presence of wave spray and white caps (8). This is normally considered as noise in methods for sensing sea surface temperatures. However, with Scanning Multi-channel Microwave Radiometer (SMMR), wind speeds can be extracted from the relative magnitudes of the radiation measured by the five channels.

Algorithms for extracting wind information from SMMR data are presently in their infancy. Methods have been proposed that use models of the radiative path through the atmosphere and account for absorption of radiation by water vapor and cloud liquid water (8, 9, 10). These algorithms are complicated and the parameters affecting absorption are not perfectly measured. Simpler empirical methods developed from statistical correlations between wind speed and the 10 channel measurements on SMMR also can be used (5 frequencies with 2 polarizations). Wind direction information is not possible from these data.

5. RADAR BACKSCATTERING ALTIMETERS

Radar altimeters have been flown on two satellites, the Geodynamic and Experimental Oceanographic Satellite (GEOS) and Seasat, for the purpose of mapping the geoid of the earth and topography of the ocean surface. The strength of the reflection of the radar pulse off the ocean surface is affected by its roughness. This is very similar to the sun glitter reflection. Methods for extracting wind speed and wave height information as by-products from the altimeter data have been developed (11, 12), and global climatologies also have been compiled (13, 14).

No wind direction information can be obtained since the radar is viewing at a 0° nadir angle (straight down).

6. MULTI-DIRECTIONAL SCATTEROMETER

By reflecting radar pulses off the ocean from two oblique directions, both the wind speed and direction can be obtained (15). This method was tested on the experimental Seasat for a short period of 3 months with good success. Wind direction information is obtained from the difference of the radar reflectivity as viewed from two antennas, one scanning forward of the satellite and a second to the aft. From this instrument, two components of the wind could be obtained. However, the signal was seldom strong enough relative to the inevitable noise to determine if the wind component was toward or away from the antenna. This produced two to four possible wind vectors for each spot on the ocean scanned by the sensor. Selecting the correct wind direction (alias removal) has been the major obstacle to the use of the data because it requires either a second data source or a synoptic interpretation of the scatterometer directional pattern. This, fortunately, is a solvable problem requiring some extra effort in the processing of the data (16). Future scatterometer plans call for the inclusion of an extra antenna which may eliminate the problem for most weather situations (17).

7. SAMPLING CHARACTERISTICS

Cloud motion data are strongly dependent on the ability to see low clouds on satellite images. Fortunately the tropical ocean is covered with low clouds over most of its vast area. An example of the coverage obtained for 10 days of FGGE is shown in Figure 1. Measurements can be made at resolutions of 100 km or less where tradewind cumulus appear; however, the continuous operationally produced measurements of NOAA/NESDIS and ESA are spaced at 250-500 km. High resolution measurements were made by Wisconsin only for the one year of FGGE which ended in November 1979. They were confined to the tropics from 15°N to 15°S in the western hemisphere (20°W-180°W) and over the Indian Ocean from 30°E to 100°E. The measurements of NOAA/NESDIS, ESA, and JMA have been continued as "operational" wind products.

Low clouds can be obscured in areas where high clouds are present, such as the Inter-Tropical Convergence Zone (ITCZ) or where cumulonimbus clouds are present. This is evident in the coverage of the "operational" cloud motions in the Pacific (Fig. 1). Dry areas where few clouds are present also have been found, such as the Arabian Sea in the spring before the development of the summer monsoon (18) and the eastern Atlantic near the north African coast when Saharan dust intrusions occur.

Wind information in cloudless areas sometimes can be supplemented with ship data, as in the case of the Arabian Sea and the Eastern Atlantic where major trade routes are present. In part of the ITCZ, little wind information can be obtained because both low cloud and ship observations are sparse. Aliased sampling is inherent in these data, and thus time averaged composites are more representative than instantaneous

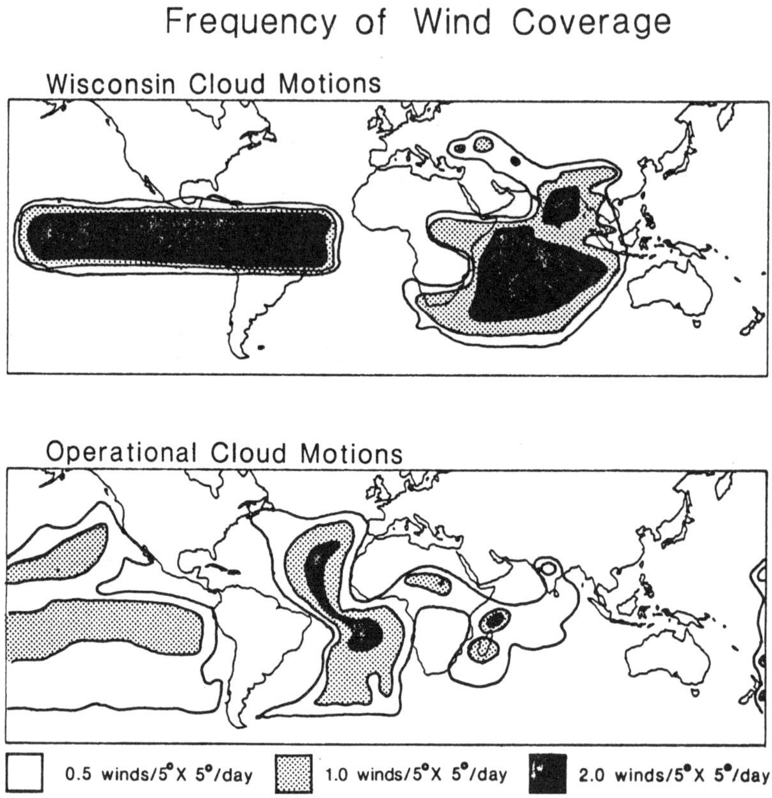

Fig. 1: The coverage of surface wind observations possible from cloud motions. Data examples were from FGGE during 10-20 July 1979.

measurements. Three to five-day composites of the cloud data were found to compare satisfactorily with wind observations (19).

Cloud motions can be obtained every hour, in principle, from the geostationary satellites; however, the cost of such intense sampling is prohibitive for continual operational programs. NOAA/NESDIS, ESA, and JMA routinely measure cloud motions two to three times/day.

Wind shear below the clouds is typically a directional turning of 16° (northern hemisphere) and a speed reduction of 25%. The turning angle decreases slightly for higher speeds, along with the speed reduction. The scatter in the ship/cloud wind comparisons reduces for the higher winds of the summer monsoon in the Indian Ocean (4, see Fig. 2), indicating that the cloud data are more representative of the surface winds for stronger (> 10 m/s) than lighter winds. Large wind directional shears (> 40°) have been found for the eastern Pacific Ocean along the southwestern coast of Mexico (20). The cloud motions are predominantly from the east, while the ship observations indicate winds from the north or south which bear a poor resemblance to the cloud winds. This decoupling of the cloud motions for the surface appears to be related to the mountains to the east and is restricted to only a small area of the North Pacific. This is also a popular merchant ship route, so cloud-ship composites made should be more valid than the cloud data by themselves.

Sun glitter wind measurements are limited to a zonal belt of approximately 15° of latitude wide, under the position of the sun with geostationary satellite data. With polar orbiting satellites, these measurements can be extended to 45° from the equator; however, the orbits are designed to minimize sun glitter coverage in the images.

The largest problem with the sun glitter method is extracting accurate measurements of the brightness of the sea surface. Satellite sensors have wide fields of view, 1 km or greater, and small clouds below the sensor's resolution cannot be detected or eliminated from the data. The surface brightness has to be selected from brightness histograms of an area, hoping that some of the pixels in the image are cloud free. Dust or aerosols present in the atmosphere also cause the same problem. Near northern Africa, sun glitter measurements are impossible during some months because of the massive clouds of dust that come from the Saharan Desert. Atmospheric contaminants increase the image brightness over the surface reflection of the sun. The brightness contributions from atmospheric scattering cannot be separated from the surface reflection in the images. The

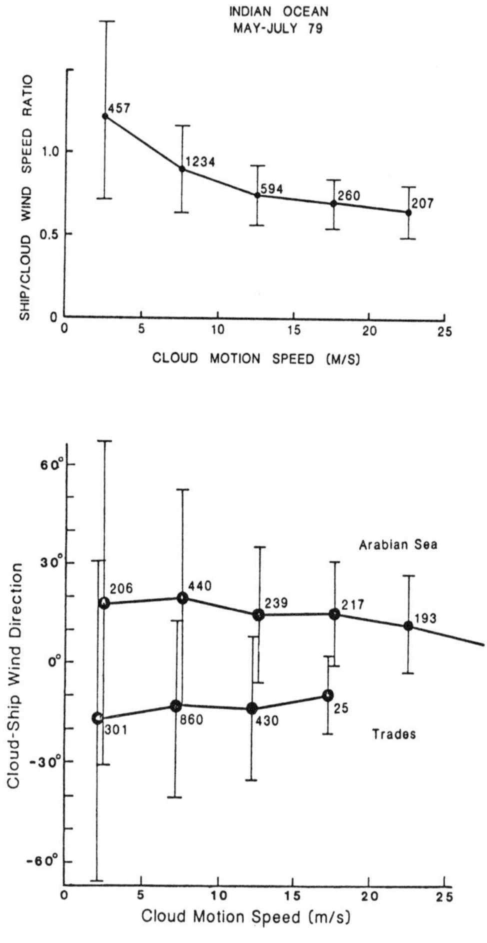

Fig. 2: The cloud-ship wind direction shear and the ship/cloud speed shear expressed as functions of the wind speed from (4). Error bars for 5 m/s speed intervals are shown.

few studies using sun glitter were made in areas where atmospheric contaminants were low and assumptions of their brightness contributions were made.

Thermal microwave and radar data have been restricted to the polar orbiting satellites because of the huge antennas required for geostationary satellites. Polar satellites imposed a sampling limit of one or less observations/day in the tropics (Fig 3.). Increased coverage is obtained in polar latitudes because of overlap between orbits. The swath width of these sensors also was restricted, so that large gaps between orbits occur in the tropics. The thermal microwave and directional radar sensors typically scan swaths from 800 to 1200 km in width. The inter-orbit gaps are usually as wide as the swath under the satellite where wind information is obtained. Full coverage of the tropics can only be obtained by compositing two or more days of orbits. The SMMR in particular is not used every day on Nimbus 7 due to its high power requirements. A diurnal cycling is employed.

The resolution of the active radars is higher than any of the other wind measuring systems. The scatterometer was digitized at a 10 km resolution on SEASAT, and the altimeter at 14 km. Wind measurements can be made at resolutions of 10-50 km for active radars, whereas the passive SMMR is applicable to a 100 km area.

The radar altimeter is further limited in that it views only straight down beneath the satellite. Wind information can only be obtained in a line along the orbit track, and thus only sparse data sampling can be obtained (Fig. 2) In the tropics, the daily spatial coverage is $\sim$1 obs./15° longitude. During the flight of GEOS-3, it was cycled on and off because of power requirements.

8. ACCURACY OF WIND MEASUREMENTS

Most wind measurements over the oceans claim observed accuracies from ±1.3 to 2.0 m/s (Table 3). These numbers can vary, depending on the quality of the "surface truth" used to evaluate them and the amount of smoothing applied. Comparisons of radar altimeter and scatterometer winds to specially instrumented ships and buoys have yielded accuracies of ±1.6 and ±1.3 m/s for each instrument respectively (21). Few comparisons have been made using winds from thermal microwave data. The authors have made comparisons with merchant ship reports averaged over 300 km around the microwave cells and found an empirical algorithm which yields a standard deviation of 1.9 m/s. Sun glitter measurements have been compared to similar synoptic data (7) and scientific buoys (22) with standard

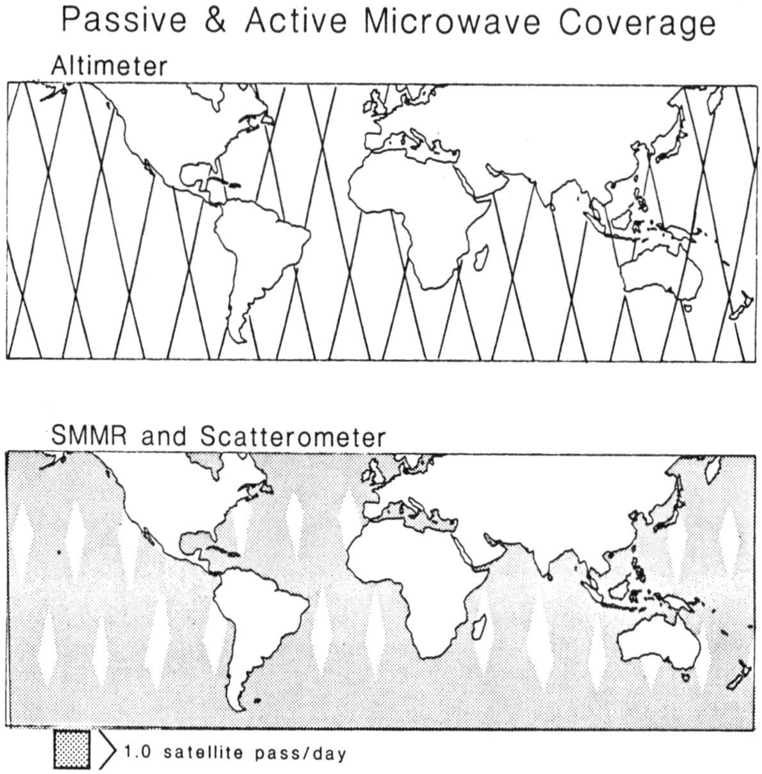

Fig. 3: An example of the one-day coverage of a polar orbiting satellite with an altimeter and a scatterometer or passive microwave sensor. This information was provided by the Jet Propulsion Laboratory, Pasadena, CA.

Table 3: Accuracy of Satellite Winds Against Surface Marine Data

Method	Speed	Direction	Range Tested
Cloud Motions	±2.1 m/s High Speeds	± 40° or ±25° for	0-24 m/s
Sun Glitter	±2.0 m/s	Not Meas.	0-15 m/s*
Thermal Microwave	±1.9 m/s	Not Meas.	0-13 m/s
Radar Altimeter	±1.6 m/s	Not Meas.	0-21 m/s
Directional Scatterometer	±1.3 m/s	±16°	0-26 m/s

* Sensitivity range has an upper limit around 15 m/s

deviations of around 2.0 m/s. Cloud motions were compared to merchant ship reports, with cloud motions extrapolated to the surface agreeing with individual ship obs. from ±2.6 to ±2.9 m/s (3, 4).

Quantitative accuracy levels are difficult to establish because of the source of "truth" data to which satellite data can be compared. Merchant ship reports, the most plentiful data source, disagree with neighboring obs. at the 2.5 to 3.5 m/s level. Even specially placed anemometers on research ships disagree by 1.7 m/s from the turbulent fluctuations in the wind (23). Comparisons of gridded analyses of ship obs. averaged over one day to cloud motions agree at the 2.0 m/s level because of the smoothing of the ship data in the analyses. The levels listed in Table 2 indicate the best information on each sensor from a variety of comparisons with different "ground truth" sources. The important points are that satellite methods are as good as or better than merchant ship reports and the active radars; the altimeter and scatterometer excel above the other passive methods.

The accuracy of the combined cloud-ship analyses produced by Hawaii and Wisconsin, however, cannot be established because all available data were used in the analyses. Their accuracies are expected to be better than each data source by itself.

Wind directions can be measured to an accuracy of 16° with the scatterometer once the correct choice of the one in four vectors is made (alias removal). Analyses made with scatterometer data have revealed mesoscale detail in winds which have seldom been seen in other oceanic analyses because of the lack of data (Fig. 4). Cloud motions have compared to ship winds with accuracies of 25° to 43° in direction. The worst case value (43°) was found for low wind speeds (less than 10 m/s). The ability of cloud motions to estimate surface wind directions improves with speed, as previously mentioned (Fig. 2).

The range over which each method can measure wind speeds is important. An upper limit on the radar altimeter and scatterometer was suspected, but scatterometer data have been collected up to 26 m/s without finding any sensitivity limits. Cloud motions also do not have any indications of speed limits.

The sun glitter, on the other hand, loses its sensitivity to winds at speeds of 13 to 15 m/s. It works best for low wind speeds. This was documented for the shape of the glitter reflection patterns studied by Cox and Munk (5).

9. PAST DATA COVERAGE

Visible satellite imagery has been taken by geostationary satellites over the United States since 1969. The first geostationary satellites covered the western Atlantic and a small part of the eastern Pacific oceans. The U. S. started running two geostationary satellites in 1975. The Meteosat and Japanese satellites were launched in 1978, and Insat was launched over the Indian Ocean in 1983.

Cloud motion measurements were initiated in the western hemisphere in 1975 by NOAA/NESDIS of the United States (Fig. 5). These data records are continuous in the western hemisphere from 1975 to the present. Cloud motion data from Meteosat and the GMS began in 1978 and were interrupted by a 1.6 yr failure of Meteosat. The Indian Ocean has a one-year record from FGGE in 1979, with a gap in coverage until the initiation of Insat in late 1983.

Sun glitter observations have never been routinely produced, but are possible on all of the geostationary satellites and also on the polar orbiting satellites which extend back to the mid 1960's. However, a systematic archive of digital geostationary satellite data dates only from 1976 onward.

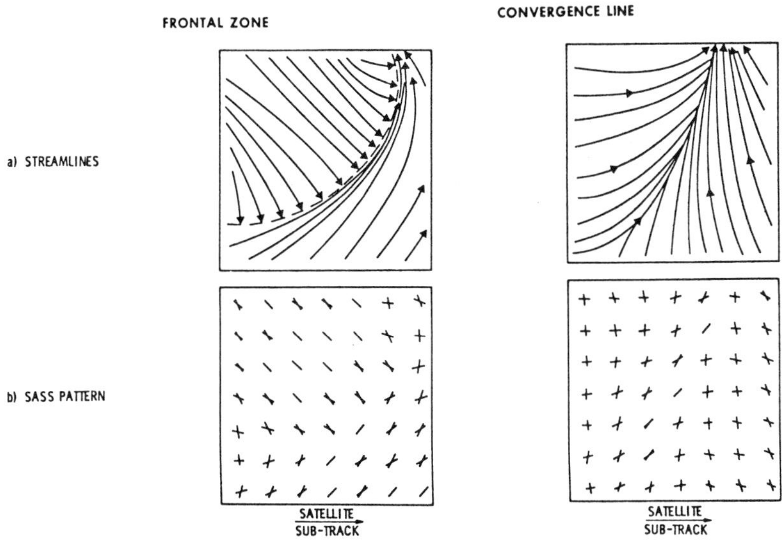

Fig. 4: An example of wind observations made from the Seasat scatterometer provided by P. Woiceshyn of J.P.L. and described in (16).

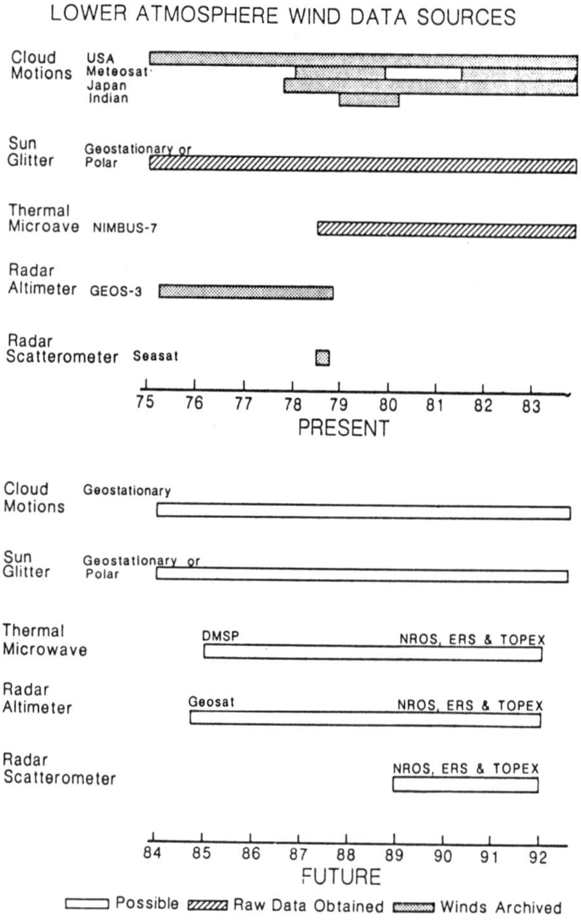

Fig. 5: The past and future of satellite sensors for measuring winds over oceans.

Thermal microwave data has been continuously collected since July of 1978. The Nimbus 7 experimental satellite has a complete record from Oct. 1978 to the present.

Radar altimeter data has been collected from April 1975 to December 1978 by GEOS-3. An overlap in coverage by two satellites occurred for three months from July to October 1978, when a similar instrument was flown on Seasat. The coverage of GEOS-3 is sporadic because of duty cycling of the sensor for power restrictions and the lack of on-board storage of data. The coverage is limited to areas of the world in view of receiving stations.

The one and only radar scatterometer flight was on Seasat for 92 days in 1978.

10. THE FUTURE

Cloud motion measurements will be continued without any planned termination. The only interruptions in these data may come from the failure of a satellite without a planned replacement such as happened with Meteosat. Changes in the design of the geostationary satellites will occur around 1990 when the present GOES series of satellites is replaced by a new design for the United States.

A thermal microwave sensor is planned for the U. S. Defense Meteorological Satellite Program (DMSP) in 1985. A gap in coverage may occur between this satellite and the present Nimbus-7 which is expected to expire in the near future. Additional thermal microwave sensors are planned for NROSS and ERS-1 which will be launched in late 1988.

A radar altimeter is planned for Geosat which will be a military satellite for the United States scheduled for launch in 1984. Present plans are for an 18-month mission; however, this may be extended for the life of the satellite, which is expected to be 5 years or more. The data also will be classified in its initial stages, and plans for distribution to scientific users have not been made at the time of the writing of this report. Other altimeter data may be available from NROSS and ERS-1 in 1988 and the proposed TOPEX.

Two radar scatterometers are planned for launch in late 1988 or 1989. The NROSS satellite will carry a scatterometer similar to Seasat (with direction improvements). ESA will launch ERS-I which will have a radar system also capable of wind sensing although it will be using a different frequency than the U. S. scatterometer and algorithms will have to be developed. These satellites are both one-mission experimental satellites. That is, no successors are planned at this time. NROSS in particular will be planned for a 18 to 36 month mission.

11. A COMPOSITE DATA SET

The main problem in planning any experiment or compiling a climatological record is the availability of the satellite data. The best wind sensor, the scatterometer, has a very short record and is not planned to fly again until late in 1988. Even if these future scatterometers are successfully deployed, their lifetime will be short, 1.5-3.0 years. The United States does not have any plans for a second NROSS, while ESA's plans for a successor to ERS-1 remain tentative. Thus, another short record of excellent data can be anticipated. The governments building these satellites should be encouraged to plan for additional flights of the same nature and collect a longer time series of these data.

The radar altimeter has a longer historical record (3.5 yrs), and will be returned to operation in the near future. This schedule is more encouraging than the scatterometer. However, Geosat has been designed for a 1.5 year geodesic mission by the U.S. Navy. Requests have been made to the Navy to extend its mission and make its data available for oceanographic uses, but a strong interest by the scientific community is needed to encourage governmental agencies to do this. Three other altimeters appear in the plans for satellites in 1988, so a long record of these data may be possible if the Geostat mission is extended.

The altimeter has two large obstacles to its use. It lacks directional measurements, making it virtually useless for momentum studies without a source of directional information. The second problem with the altimeter is its narrow swath coverage, a line of observations only directly under the spacecraft. This implies that synoptic coverage of an ocean basin can only be obtained by compositing several days of data. Some temporal information must be compromised.

Cloud motions have a long history and the best prospect for continued coverage in the future. They are obtained from a popular sensor which has a wide variety of uses -- the imaging scanners on geostationary satellites. The sponsors of these satellites plan to continue their operation because it can be done with proven and tested technology, and the uses of image data for weather forecasting are well known.

Cloud motions are a source of surface data inferior to the radar sensors. This is because of the variability of the atmospheric boundary layer below the clouds and the dependence on finding low level clouds on satellite images. But they are our only continual source of information until a better sensor is in operation.

The other methods, thermal microwave and sun glitter, offer some information on surface wind speeds. They have not been extensively tested, so their accuracies are not fully developed. Methods of applying them to a wide range of conditions over the full earth also have to be established.

Three methods, microwave, altimeter, and sun glitter, do not measure wind direction. The scatterometer can provide it with some additional information, and may be able to do so by itself in the future. This is the value of cloud motion data. A combination of data sources is certainly possible where the strengths of each are exploited. The instruments that are sensitive to surface roughness can give speed information and wind stress magnitude. Wind direction can be obtained from cloud motions.

The cloud motions also can provide data in the gaps between orbits of the polar satellites. The subcloud wind shear statistics used to adjust cloud motions to surface winds, could be fine tuned by comparisons with surface measurements where cloud and polar orbitor data overlap. The data from radar altimeters in particular, are far more accurate than the ship reports. Seasonal climatologies of the wind shear have had to be used because of the scatter in the ship reports. With more accurate data from a radar altimeter, changes in the vertical wind shear on shorter time scales could be followed. It may be possible to update the wind shear algorithm for each cloud motion data set; or one could treat the vertical shear as an important parameter in its own right, having direct relationships to air-sea fluxes of heat and moisture. The possibilities of this data combination should be explored.

12. REFERENCES

1. Sadler, J.C. and Kilonsky, B.J., 1981, Trade wind monitoring using satellite observations. Dept. of Meteor. Rept. UMET81-01, Univ. of Hawaii, 72 pp.

2. Hasler, A.F., Skillman, W.C., Shenk, W.C., and Steranka, J., 1979, In situ aircraft verification of the quality of satellite cloud winds over oceanic regions. J. Appl. Meteor. 18, 1481-1489.

3. Wylie, D.P., and Hinton, B.B., 1981, The feasibility of estimating large-scale surface wind fields for the summer MONEX using cloud motion and ship data. Bound. Layer Meteor., 21, 357-367.

4. Wylie, D.P., and Hinton, B.B., 1982, A comparison of cloud motion and ship wind observations over the Indian Ocean for the year of FGGE. Bound. Layer Meteor. 23, 197-208.

5. Cox, C., and Munk, W., 1954, Measurements of the roughness of the sea surface from photographs of the sun's glitter. J. Opt. Soc. Amer. 44, 838-850.

6. Levanon, N., 1971, Determination of the sea surface slope distribution and wind velocity using sun glitter viewed from a synchronous satellite. J. Phys. Oceanogr. 3, 214-220.

7. Wald, L., and Monget, J.M., 1983, Sea surface winds from sun glitter observations. J. Geophy. Res. 88, 2547-2555.

8. Wilheit, T.T., and Chang, A.T.C., 1980, An algorithm for retrieval of ocean surface and atmospheric parameters from observations of the scanning multichannel microwave radiometer. Radio Sci. 15, 525-544.

9. Wentz, F.J., V.J. Cardone, and Fedor, L.S., 1982, Intercomparison of wind speeds inferred by the SASS, altimeter, and SMMR. J. Geophy. Res. 87, 3378-3384.

10. Cardone, V., Chester, T., and Lipes, R., 1983, Evaluation of Seasat SMMR wind speed measurements. J. Geophy. Res. 88, 1709-1726.

11. Fedor, L. S., and G.S. Brown, 1982, Wave height and wind speed measurements from the Seasat Radar Altimeter. J. Geophy. Res. 87, 3254-3260.

12. Brown, G.S., Stanley, H.R., and Roy, N.A., 1981, The wind speed measurement capabilities of spaceborne radar altimetry. IEEE J. Oceanic. Eng. OE-6, 59-63.

13. McMillan, J.D., 1981, A global atlas of GEOS-3 significant wave height data and comparison of the data with national buoy data, NASA CR 156882, Wallops Center.

14. Chelton, D.B., Hussey, K.J., and Parke, M.E., July, 1981, Global satellite measurements of water vapor, wind speed, and wave height. Nature.

15. Schroeder, L.C., Boggs, D.H., Dome, G., Halberstram, I.M., Jones, L.W., Pierson, W.J., and Wentz, F.J., 1982, The relationship between wind vector and normalized radar cross section used to derive Seasat-A satellite scatterometer winds. J. Geophy. Res. 87, 3318-3336.

16. Wurtele, M.G., Woiceshyn, P.M., Peteherych, S., Borowski, M., and Appley, W.S., 1982, Wind direction alias removal studies of Seasat scatterometer-derived wind fields. J. Geophy. Res. 87, 3365-3377.

17. Schroeder, L.C., Grantham, W.L., Bracalente, E.M., Britt, C.L., Shanmugan, K.S., Wentz, F.J., Wylie, D.P., and Hinton, B.B., 1983, A study of removal of ambiguous wind direction results for a Ka-band scatterometer wind sensor using measurements at three different azimuth angles. International Geoscience and Remote Sensing Symposium, San Francisco, CA, September 1983.

18. Wylie, D.P., and Hinton, B.B., 1982, The wind stress patterns over the Indian Ocean during the summer monsoon of 1979. J. Phys. Ocean. 12, 186-189.

19. Halpern, D., and Knox, R.A., 1983, Coherence between low level cloud motion vectors and surface wind measurements near 0°, 152°W from April 1979-February 1980. Atmosphere-Ocean.

20. Wylie, D.P., Hinton, B.B., and Howland, M., 1983, Surface wind observations from cloud motions during FGGE. Tropical Ocean-Atm. News Letter 18 (May), 2-4.

21. Lame, D.B., and Born, G.H., 1982, Seasat measurement system evaluation: achievements and limitations. J. Geophy. Res. 87, 3175-3278.

22. Wylie, D.P., Hinton, B.B., and Millett, K.G., 1981, A comparison of three satellite based methods for estimating surface winds over oceans. J. Appl. Meteor. 20, 439-449.

23. Pierson, W.J., Peteherych, S., and Wilkerson, J.C., 1980, The winds of the comparison data set for the Seasat Gulf of Alaska Experiment. IEEE J. of Oceanic Eng. OE-5, 169-176.

13. ACRONYMS AND SATELLITES

ERS - Earth Resources Satellite

ESA - European Space Agency (International)

FGGE - First GARP Global Experiment (International, from December 1978 through November 1979)

GARP - Global Atmospheric Research Program (International)

GEOS - Geodynamic and Experimental Oceanographic Satellite (U.S.)

Geosat - A geodynamic satellite (U.S.)

GMS - Geostationary Meteorological Satellite (Japan)

GOES - Geostationary Operational Environmental Satellite (U.S.)

JMA - Japanese Meteorological Agency (Japan)

Insat - A geostationary satellite over the Indian Ocean (India)

Meteosat - A European geostationary satellite (International)

Nimbus-7 - An experimental Polar Orbiting Satellite (U.S.)

NOAA/NESDIS - National Oceanographic and Atmospheric Administration/National Environmental Satellite and Data Information Service (U.S.)

NROSS - Navy Remote Ocean Sensing System (U.S.)

Seasat - An experimental oceanographic satellite (U.S.)

SMMR - Scanning Multi-frequency Microwave Radiometer (U.S.)

LARGE SCALE SURFACE HEAT FLUXES

E.S. Sarachik

Center for Earth and Planetary Physics
Harvard University
Cambridge, MA 02138

ABSTRACT

The heat flux through the ocean surface, Q, is the sum of the net radiation at the surface, the latent heat flux into the atmosphere and the sensible heat flux into the atmosphere (all fluxes positive upwards). A review will be presented of the geographical distribution of Q and its constituents, and the current accuracy of measuring Q by ground based measurements (both directly and by "bulk formulae") will be assessed.

The relation of Q to changes of oceanic heat content, heat flux and SST will be examined and for each of these processes, the accuracy needed for Q will be discussed. The needed accuracy for Q varies from process to process, varies geographically, and varies with the time and space scale considered.

I. INTRODUCTION

Interchanges of heat and momentum at the ocean surface are crucial determinants of the state of the earth's coupled ocean-atmosphere system. Water vapor evaporated at the ocean surface is the dominant source of precipitation over both land and ocean. The flux balance at the ocean surface helps determine its temperature and in turn the sea surface temperature helps determine the forcing of both atmospheric and oceanic motion.

Measurement of these fluxes has been accomplished largely inferentially, through the gathering of large volumes of shipboard data over the last 60 or so years. This data has been

C. Gautier and M. Fieux (eds.), Large-Scale Oceanographic Experiments and Satellites, 147–165.

adequate to describe the mean state and seasonal variability of the fluxes but vast geographical gaps, especially in the Western Pacific and the Southern oceans, still exist. The gradual emergence of satellite techniques to measure these fluxes for the first time provides the possibility of a synoptic picture of heat fluxes at the ocean surface but it is an item of obvious truth that the calibration of satellite measurements can be no better than the directly measured surface quantities.

It is the purpose of this paper to review in broad strokes what we know about fluxes at the ocean surface, to describe how well we know them from direct measurements and other ground based inferences, and then to consider how well we have to know them. The accuracy required for these fluxes varies with the physical process involved, with the geographical location, and with the space and time scale considered.

The starting point for the discussion of large scale heat fluxes is the heat balance at the ocean surface (all fluxes are defined to be positive *upwards*).

$$-F_{net} = LE + S - Q \qquad (1)$$

where

LE = The latent heat flux
S = Sensible heat flux into the atmosphere
Q = Sensible heat flux into the ocean

and

F_{net} = Net radiative flux at the surface.

The latent heat flux is the product of the latent heat of evaporation (or condensation), L, and the evaporation rate. The units are such that 1 mm/day of evaporation corresponds to a latent heat flux of 29 W/m^2.

The net radiative flux has two parts:

$$F_{net} = F_{\odot}^{net} + F_{ir}^{net}$$

where the net solar flux at the surface is $F_{\odot}^{net} = (1-\alpha)F_{\odot}$ where $F_{\odot}$ is the actual solar flux hitting the surface (the amount that survives reflection from clouds and aerosols and absorption in the interior of the atmosphere) and α is the albedo of the surface of the ocean, generally only a few percent (note that by our sign convention the solar flux is negative).

The net infrared flux is composed of two terms:

$$F_{ir}^{net} = \varepsilon\sigma T_s^4 + F_{ir}^{\downarrow}$$

where the first is blackbody emission from the surface of the ocean with surface temperature T_s and emissivity ε (about .97), and the second is the downward infrared flux due to emission from the gaseous and particulate constituents of the atmosphere (including the emission from cloud bottoms).

To give a scale for the various numbers, we consider the globally averaged, annually averaged values of the various constituents of eq. (1).

The globally averaged latent heat flux, LE, from the surface is about 80 W/m^2 corresponding to an evaporation rate of 1 m/year or 2.7 mm/day. Of this 13% is due to evaporation from the land surface, and 87% is due to evaporation from the surface of the ocean.

The globally averaged sensible heat flux, S, is about 16 W/m^2 of which 60% is from the land and 40% is from the ocean. The globally averaged Bowen ratio $B = S/LE$ is .3 with the land having $B = 1$ and the ocean having $B = .11$.

The globally averaged net radiation at the surface, F_{net}, is about 100 W/m^2.

The globally averaged sensible heat flux into the ocean is zero since on the average, the ocean is neither heated nor cooled on an annual basis.

We see that about 100 W/m^2 of sensible and latent heat enter the atmosphere. Since all the latent heat eventually condenses, the net heating of the atmosphere due to surface fluxes of 100 W/m^2 is about 1 K/day. (There is a net infrared cooling of the global atmosphere of about 1.5 K/day, the difference corresponding to the heating due to direct absorption of solar radiation in the atmosphere.) To fix magnitudes, we note that while 100 W/m^2 into the atmosphere will heat the atmospheric column by 1 K/day, 100 W/m^2 into the ocean will heat a 100 m column of water by .7 K/month.

To give an idea of the latitudinal distribution of these quantities, Fig. 1 shows the net annually averaged components of the heat budget which have been zonally averaged over the ocean surface only. We see that on the average the ocean is heated ($Q < 0$) in tropical regions and cooled in mid-latitudes by heat fluxes through the ocean surface. Figure 2 shows the seasonal variation of the zonally averaged heat flux $-\langle Q \rangle$ and indicates that in tropical regions, the seasonal variation is less than 40 W/m^2 while in mid-latitudes the seasonal variation can be an order of magnitude larger.

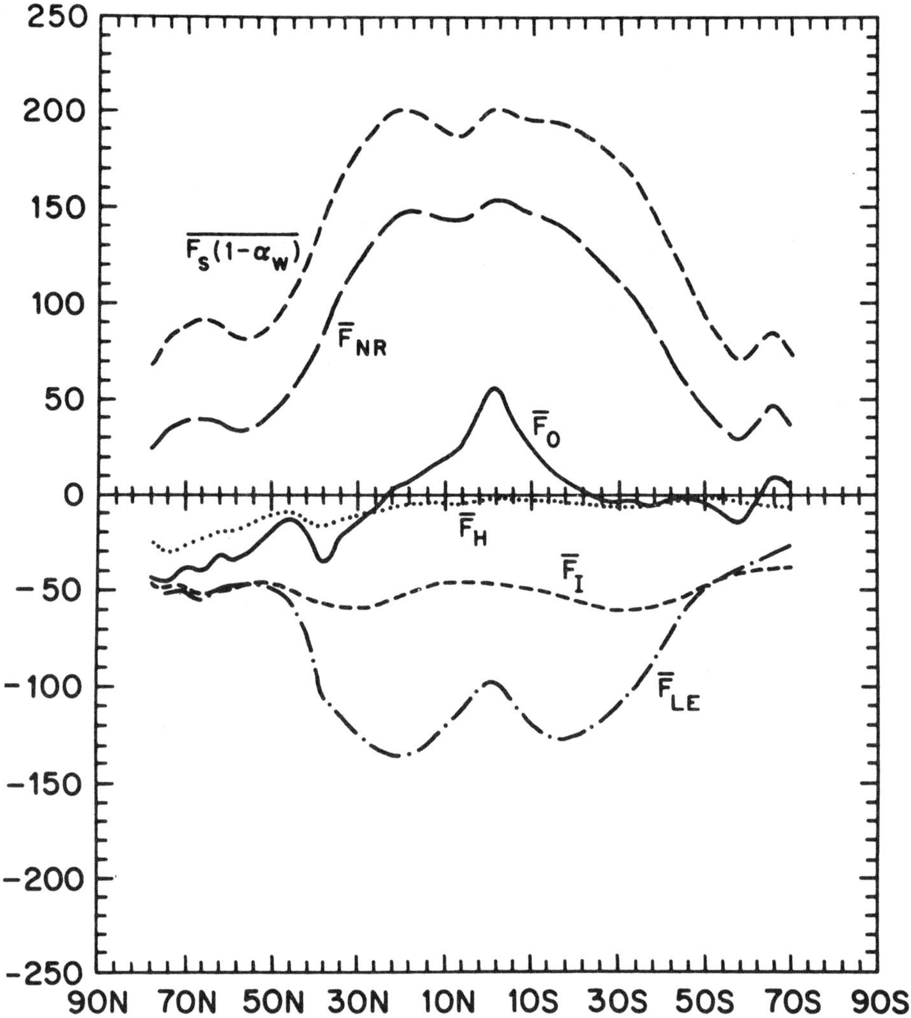

Figure 1. The latitudinal distribution of the annually and zonally averaged ocean surface fluxes. $\bar{F}_{NR}$, $\bar{F}_{LE}$, $\bar{F}_H$, and $\bar{F}_O$ correspond to our $-\langle F_{net}\rangle$, $-\langle LE\rangle$, $-\langle S\rangle$, and $-\langle Q\rangle$ respectively. The ordinate is in W/m^2. (After Esbensen and Kushnir, 1981.)

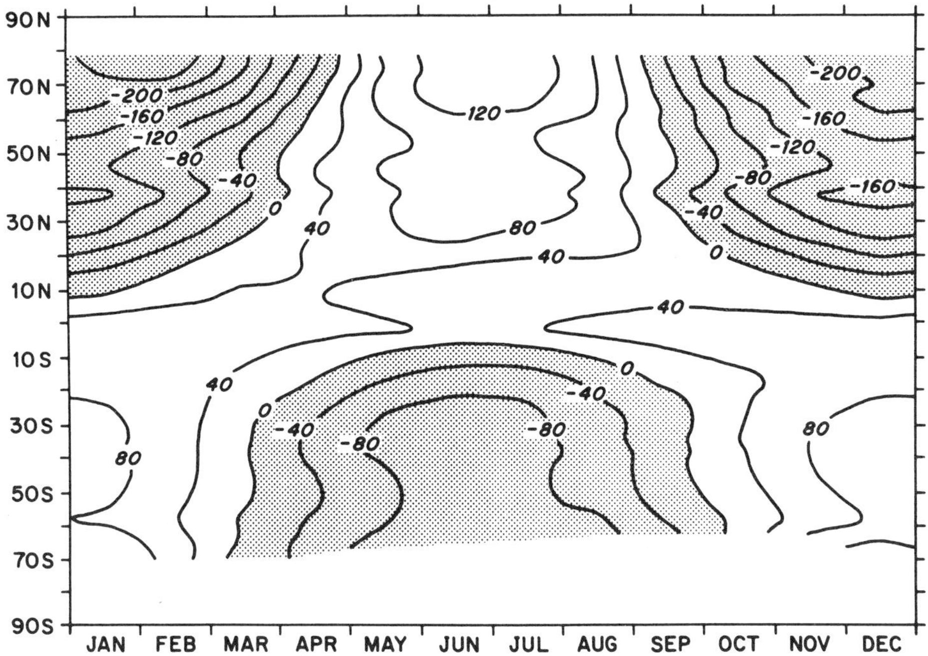

Figure 2. Seasonal variation of the zonally averaged heat flux into the ocean -<Q> in W/m^2. (After Esbensen and Kushnir, 1981.)

II. GEOGRAPHICAL DISTRIBUTION OF THE FLUXES

There have been only two global compilations of the various fluxes at the ocean surface: That of Budyko (1974) and that of Esbensen and Kushnir (1981). While these compilations differ in detail, we will concentrate on places that the various fluxes are very large or very small: in these regions the two compilations agree. We show Budyko's pictures--it has the advantage of giving land values so that the difference between land and ocean is made especially clear.

Figure 3 shows the annual solar flux at the surface. The solar flux at the surface is primarily determined by the average fractional cloudiness in the overlying atmosphere. We see that the largest values of solar insolation at the surface; values exceeding 250 W/m^2 occur over the subtropical desert regions, where cloudiness is very small. In mid-latitude regions, winter cloudiness tends to be greater than summer cloudiness and the annual average tends to be dominated by the winter cloudiness.

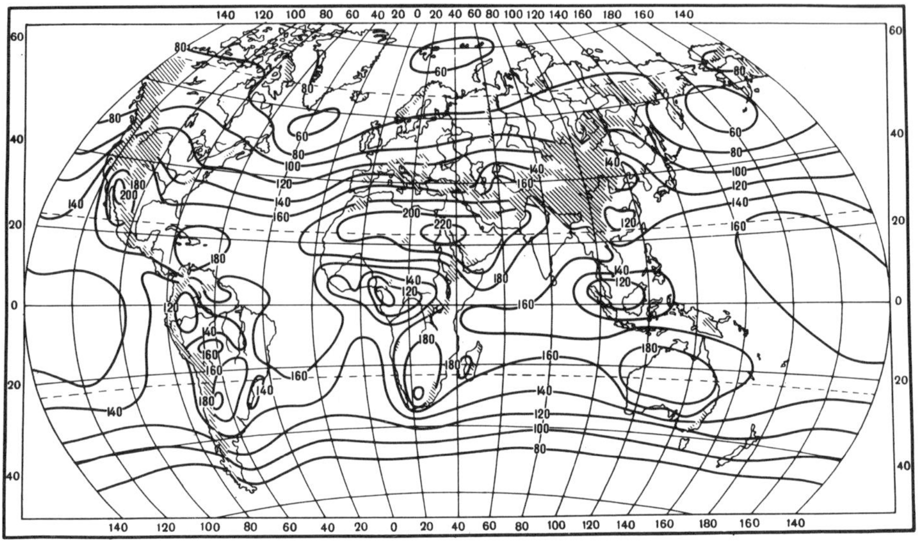

Figure 3. Net solar flux at the surface in Rcal $cm^{-2}yr^{-1}$. Note that 100 kcal $cm^{-2}yr^{-1}$ = 133 W/m^2 (after Budyko, 1974).

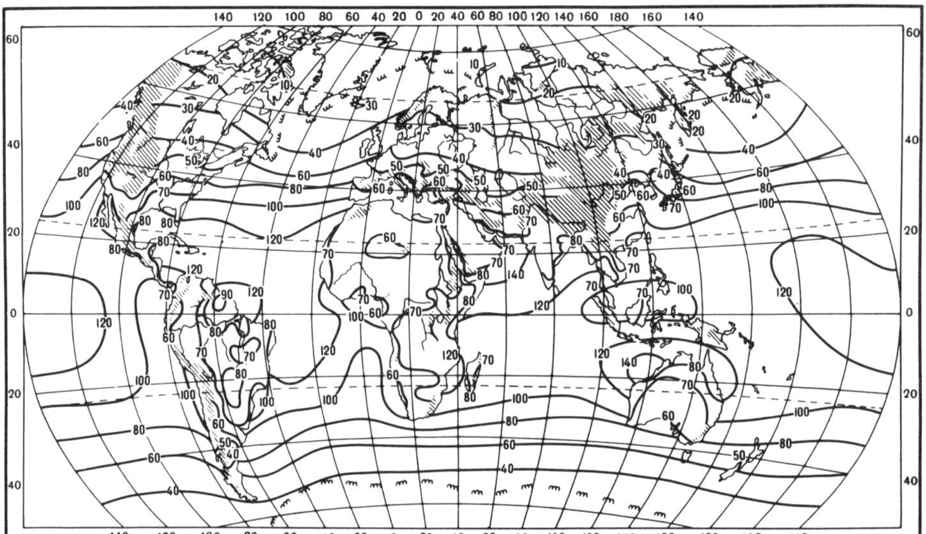

Figure 4. Net radiation flux at the surface. Units as in Fig.3.

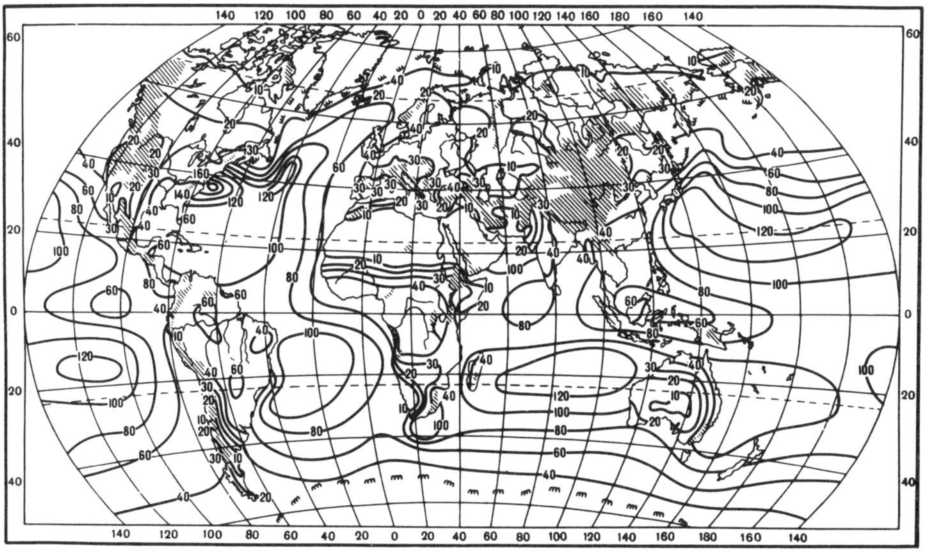

Figure 5. Latent heat flux into the atmosphere. Units as in Fig. 3. 100 kcal $cm^{-2}yr^{-1}$ corresponds to 4.6 mm/day of evaporation. (After Budyko, 1974.)

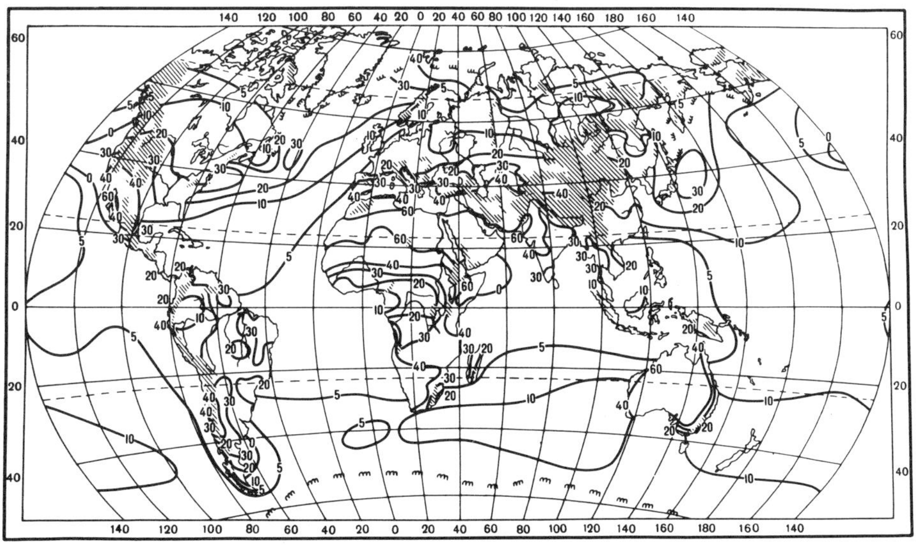

Figure 6. Sensible heat flux into the atmosphere. Units as in Fig. 3. (After Budyko, 1974).

In particular, the solar flux at the ground decreases poleward. The flux is relatively large over the tropical oceans where winter cloudiness is not much greater than summer cloudiness with values of 200 W/m^2 being typical. Finally note that the solar flux at the surface is relatively small over regions of persistent precipitation and cloudiness, e.g., over the maritime continent separating the east Indian Ocean from the Western Pacific.

Figure 4 shows the net radiation at the surface. F_{net} is large when either the net solar flux is large or the net outgoing radiation is small. The largest values occur over tropical oceans with values of 150 W/m^2 being typical.

Figure 5 shows the latent flux from the surface (recall that 29 W/m^2 corresponds to 1 mm/day of evaporation). The latent heat flux is limited by the net radiation at the surface and depends on a moisture source. It is therefore large over tropical oceans, especially in mid oceans where the water is warm and cloudiness is not large. The largest regional values occur over the Gulf stream where warm water supplies moisture to intense wintertime storminess in the Eastern Atlantic.

Figure 6 shows the sensible heat flux into the atmosphere. It tends to be small over the oceans except in the regions of western boundary currents where warm water flows beneath cold winter air, facilitating large transfers of heat to the atmosphere.

Figure 7 shows the heat flux into the ocean. As we saw in Fig. 1, heat goes into the ocean, in tropical latitudes and is returned to the atmosphere at higher latitudes. We see in Fig. 7 that heat enters the tropical ocean where it is, on the average, relatively cool. Thus there are large heat fluxes into the ocean on the eastern flank of the Atlantic and Pacific and on the western flank of the Indian Ocean. Heat leaves the ocean in mid-latitude where it is relatively warm, especially in the regions at the western boundary currents and their mid-ocean extensions. This way of looking at the fluxes is slightly dangerous because much of the heat entering the tropical oceans occurs not where the flux is largest, but over broad reaches of the tropical mid-oceans where the fluxes are moderate (e.g., 30 W/m^2). As we will see in Section III, it is precisely these moderate values that are most subject to error, yet it is these regions of moderate input which determine the net heat transport of the ocean (Section IV).

III. OUR CURRENT MEASUREMENT CAPABILITY

The development of accurate satellite techniques for measuring the components F_{net}, LE, and S of the surface heat budget (1) will depend crucially on a capability for calibrating the satellites

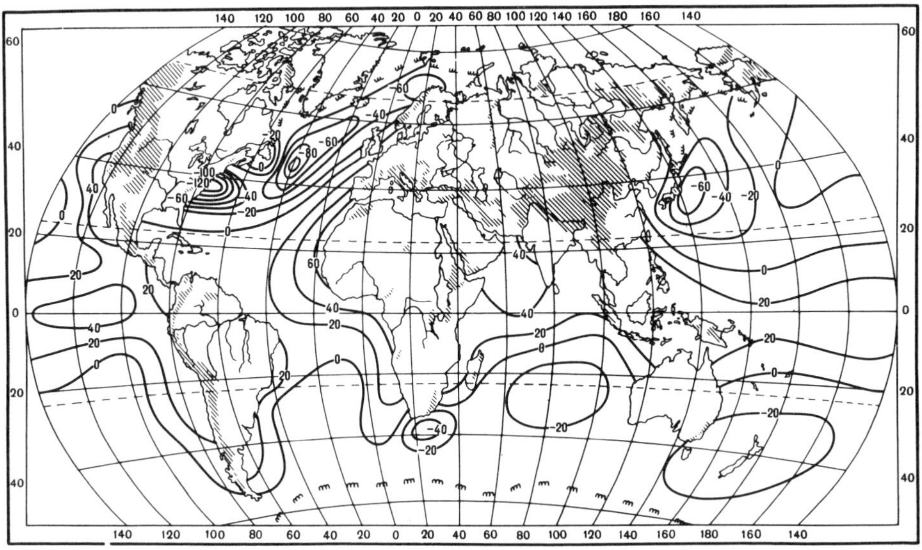

Figure 7. Heat flux into the ocean. Units as in Fig. 3. (After Budyko, 1974.)

with ground based measurements. It is clear that satellite measurements can be no more accurate than the ground based measurements by which they are calibrated. It is also clear that even when calibrated against inaccurate ground based measurements, the increase in geographical average and time sampling that satellites provide will make satellite measurements extremely valuable. An assessment of our current ability to measure fluxes at the ocean surface seems a necessary prerequisite to understanding the potential impact of satellite measurements.

There are two types of measurements for ocean surface fluxes --direct and indirect. The direct measurement is a *point* measurement. The indirect measurement (bulk formulae) is generally calibrated at a point against direct measurements and then spatially and temporally averaged to get large scale ocean surface fluxes. *All* measurement capability eventually depends crucially on the accuracy of the direct measurement against which they are calibrated. Accuracies are defined by the scatter that successive measurements, under roughly similar conditions, undergo. Since the direct measurement is *by definition* the standard, this irreducible scatter *defines* the accuracy with which the mean can be estimated. It should be noted that direct measurements are difficult, time consuming, and expensive and the ensembles of measurements needed to give an estimate of the mean are almost never performed. We shall therefore be entered with quoting the accuracy gotten directly from the scatter. As we will see below, the situation is less good than is generally believed.

The net radiation at the surface can be measured to about 5% using an all-wave radiometer (Simpson and Paulson, 1979). Each component of F_{net}, $F_{\odot}^{net}$ and F_{ir}^{net}, can similarly be measured to 5% separately but for the purpose of the fluxes appearing in eq. (1), there is little point in doing so.

Sensible heat can be measured either by direct correlation $S = \rho c_p \overline{w'T'}$ or by the dissipation method, which consists of using the steady variance equation:

$$\frac{1}{2}\overline{T'^2} = -\overline{w'T'}\frac{\partial \bar{T}}{\partial z} + \frac{1}{2}\frac{\partial}{\partial z}\overline{w'T'^2}$$

and neglecting the final convergence term. In either case, accuracies are no better than 20% (Large and Pond, 1982). Since sensible heat over vast areas of the ocean forms a relatively small part of the surface budget, this is not too disturbing.

The latent heat can again be measured either by direct correlation $E = \rho \overline{q'w'}$ or by the dissipation method:

$$\frac{1}{2}\overline{q'^2} = -\overline{w'q'}\frac{\partial \bar{q}}{\partial z} + \frac{1}{2}\frac{\partial}{\partial z}\overline{w'T'^2}$$

again neglecting the final convergence term.

Even under the most carefully controlled conditions, the accuracy of the measurements are no better than 20% (Large and Pond, 1982). Since there are no direct measurements of Q, the sensible heat flux into the ocean, and since Q tends to be smaller than either LE or F_{net}, i.e., they tend to cancel, point determinations of the heat flux into the ocean gotten as a residual of directly measured quantities in eq. (1) would generally show inaccuracies far exceeding 20%.

It should be emphasized that the above numbers are for carefully attended special experiments and therefore represent the state of the art in what can be measured directly. Unattended instruments, with their problems of shielding, cleanliness, degradation, etc. could not be expected to yield accuracies as good as those quoted.

The vast majority of our information about surface heat fluxes, including all the diagrams shown in Section II, comes from *indirect* measurements using standard ship data for wind speed $|\underset{\sim}{u}|$, sea surface temperature T_s, fractional cloud cover n, and anemometer level temperature, T_a, and specific humidity q_a. These are used in "bulk formulae":

$$F_{\odot}^{net} = (1-\alpha) \text{ "Book value" } (1-C_0 n) \quad , \tag{2a}$$

$$F_{ir}^{net} = \varepsilon\sigma T_s^4 (a - bq_a^r)(1 - C_{ir} n) \quad , \tag{2b}$$

$$S = \rho C_p C_s |\underset{\sim}{u}| (T_s - T_a) \tag{2c}$$

and

$$LE = \rho L C_E |\underset{\sim}{u}| (q_{sat}(T_s) - q_a) \tag{2d}$$

C_0, C_{ir}, a, b, and r are constants; C_S and C_E are drag coefficients which may depend on the wind speed and atmospheric stability, and $q_{sat}(T_s)$ is the saturation value of the specific humidity at the temperature of the sea surface.

The bulk formulae in Eq. (2) are empirical: They are calibrated by regressing these formulae against as high quality measurements as can be taken on shipboard. These formulae suffer from two basic problems: calibration and data.

1. Calibration are taken against direct measurements which are themselves subject to the errors described above. Furthermore the calibrations are performed during reasonably short periods of intensive measurements. Fung, Harrison and Lacis (1983), for example, have shown, by comparison with accurate radiative transfer calculations, that the long wave radiation at the surface can vary as much as 20-30 W/m^2 by varying the *upper level* profiles of temperature and moisture over a standard deviation of these variables. Calibrations taken over periods short compared to the naturally variability of the upper air profiles will be biased. The only way to avoid this bias is to take calibration measurements over time periods long compared to the natural variability. In the tropics we may expect this to be 4-5 days and in mid-latitudes, a week. (Such long calibrations have not, to my knowledge, yet been performed, nor have any calibrations been performed in the tropics. This may account for the large number of different long wave radiation formulas in the literature, each characterized by different values of a, b, r and C_{ir} in eq. (2b).)

An additional problem with the calibrations is that they are taken at fixed geographical locations and then assumed to hold everywhere. Since each of the formulae in eq. (2) attempt to represent a complicated process depending on a large number of variables by a simple formula containing only a few surface variables, it is likely that the representation will be geographically and temporally dependent. Most calibrations have been carried out near mid-latitude weather ships and the applicability of such calibrations to tropical regions, say, remains undemonstrated.

2. The data itself, taken from routine ship measurements, is subject to inaccuracies. Typical ship errors are: 1 K in T_s and T_a, 2 m/sec in $|\underset{\sim}{u}|$, 3 gm/kg in $q_{sat}(T_s) - q_a$ and an octa or two in n. Since these errors are presumably unbiased, large amounts of data would alleviate the problem. Ship data, however, is notably inhomogeneous in space: There are large amounts of data over well-defined tracks leaving vast amounts of the ocean with little or no data. A region with only one or two ship reports per month would therefore have errors in the monthly means as large as the measurement error itself. Since in the tropics, say, $q_{sat}(T_s) - q_a$ is rarely as large as 3gm/kg, evaporation deduced from even a perfect bulk formula of type (2d) would have an error as large as the mean.

The sensible heat flux into the ocean, Q, can only be measured as a residual in eq. (1) and therefore is subject to all the inaccuracies of all the bulk formulae in eq. (2). It is probably not too pessimistic to note that monthly mean values of Q over $5° \times 5°$ squares with only a few ship observations probably has errors as large as the mean. Climatological monthly means (i.e., average of the same month over many years) presumably reduce this error but since the absolute calibration of the bulk formulae remains unknown it is difficult to estimate the error.

To summarize this section, we may say that the state of art in both the direct measurements and in the calibration of bulk formulae indicates that much more work has to be done. Long calibration experiments at widely disparate geographical locations are a necessity if we are to have any confidence in the bulk formulae. Similar calibrations of satellite data with direct measurements seems an obvious corollary.

IV. THE ACCURACY REQUIRED FOR Q

In this section, we will examine some oceanographic processes in order to determine the importance of heat fluxes into the ocean, Q, for these processes. It is the simple premise of this section that those processes for which Q is important require a greater measurement accuracy than those processes for which it is unimportant. While this premise seems obvious, the state of our knowledge of large scale oceanographic processes is such that only recently has it become possible to provide an outline of when Q is or is not important. The problem is complicated by the geographical diversity of oceanographic dynamics: Thus Q will be found to be important for a specific process in some places, but not in others.

In this section, we will examine three large scale oceanographic phenomena: The mean zonally averaged meridional heat flux, the seasonal variation of the oceanic heat content, and the variability of sea surface temperature. As we will see, only in the case of mean meridional heat flux does the *dynamics* demand that Q has to be important. For the other two processes, we will appeal to a variety of theoretical and observational arguments.

We begin by writing the heat budget of a column of unit area of ocean extending from the bottom to the surface:

$$-Q = \nabla \cdot \underset{\sim}{T} + R \qquad (3)$$

where $\underset{\sim}{T}$ is the vertical integral of the oceanic heat flux and R is the rate of storage of heat (i.e., the rate of change of the heat content),

$$R = \frac{\partial}{\partial t} \int_{bot}^{sfc} \rho c_p T \, dz$$

Equation (3) simply states that the heat flux Q entering the oceanic column is either stored (by raising the temperature) or diverged away by oceanic motions. The accuracy required for Q depends on the importance of Q in eq. (3). This will depend on the process considered, on the time scale involved, on the scale of spatial averaging, and on the geographical location.

a. The mean zonally averaged meridional heat flux

Let $\langle\ \rangle$ represent zonal average, and $\overline{}$ represent time average. The heat budget then becomes

$$-\langle\bar{Q}\rangle = \frac{d}{dy} \langle\underset{\sim}{\bar{T}}\rangle \qquad (4)$$

since in the long term average, the averaged storage rate must vanish. Ever since the pioneering work of Vonder Haar and Oort (1973) showed that the mean zonally averaged oceanic heat flux exceeds the atmospheric heat flux equatorward of 30°, it has been realized that the oceanic heat flux $\langle\underset{\sim}{T}\rangle$ is of fundamental climatic importance. The latitude integration of (4) gives

$$<\bar{\underset{\sim}{T}}> = \int_{\text{north pole}}^{y} <\bar{Q}>\, dy$$

so that not only does $<\bar{\underset{\sim}{T}}>$ depends directly on $<\bar{Q}>$, but also the errors in $<\bar{\underset{\sim}{T}}>$ accumulate with the latitude integration of the errors in $<\bar{Q}>$. Holding $<\bar{Q}>$ to a constant fractional error would produce the same fractional error in $<\bar{\underset{\sim}{T}}>$.

b. Seasonal variability of heat content and storage rate

In mid-latitudes (poleward of 15°), on large scales (thousands of kilometers) and away from boundaries, Gill and Niiler (1973) have argued, on the basis of scaling arguments, that the large scale heat transports must be very sluggish on seasonal time scales so that eq. (3) simply becomes $Q = R$. Here, of course, the accuracy demanded for Q is equal to the accuracy needed for R.

Lamb and Bunker (1982) however find in an observational study, that for zonal averages in the mid-latitudes Atlantic, the seasonal oceanic heat transport variations are of the same order as the seasonal variations of Q and R. There is no necessary contradiction between these results of the heat transport variations of Lamb and Bunker's. Zonal averages include the western boundary current where Gill and Niiler's arguments don't apply. To my knowledge, no conclusive observational test of $Q = R$ over the large regions for which Gill and Niiler's arguments should apply has been performed since the publication of their paper.

In any case, it is clear that for mid-latitude seasonal storage variations, Q is of first order importance and the accuracy required for Q is of the order of the accuracy needed for R.

The situation in equatorial regions is completely different. The upper part of figure 8, from Merle (1980), shows the storage rate R in the Western Atlantic vs. the heat flux through the surface Q averaged over a band from 6 S to 6 N. We see that the seasonal peak to peak variations in R is over 500 W/m^2 where for Q it is less than 50 W/m^2. Furthermore, Q is out of phase with R in the sense that maximum heating of the ocean by Q corresponds to minimum storage rate. Clearly in the Western Atlantic, the fundamental balance is $R = \nabla \cdot \underset{\sim}{T}$ so that storage rate (and heat content variations) are predominantly caused by internal convergences and divergences of oceanic heat flux. One might suppose that most of the oceanic heat flux is east-west: Simply sloshing heat from west to east in phase with the zonal wind variations on the equator. The lower part of figure 8, however, shows that even for zonal bands across the *entire* equatorial Atlantic, a similar conclusion holds, indicating the presence of large seasonal *meridional* transports of heat into and out of zonal bands. We may conclude that over large scale equatorial regions, improving the accuracy of Q will not help understand heat content variations. Conversely, large inaccuracies in Q won't be fatal to understanding heat content variations.

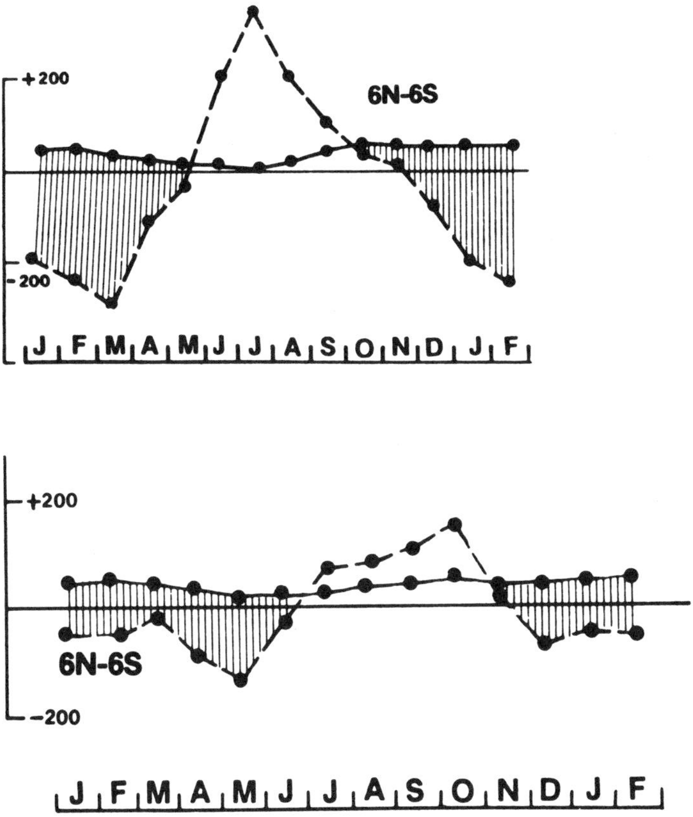

Figure 8. Seasonal variation of the storage rate (broken line) and heat flux into the ocean (solid line). Upper diagram corresponds to the Western Atlantic while lower diagram corresponds to zonal average across entire Atlantic (after Merle, 1980).

c. Sea-surface temperature variations

The sea surface temperature (SST) is the one quantity that the atmosphere and the ocean share and therefore the oceanic quantity of greatest importance for the atmosphere. The source of the difficulty in understanding SST is precisely the interactive nature of the atmosphere and the ocean. To a large extent, the SST is hardly determined by the ocean at all. To see this, imagine the atmosphere and ocean have no horizontal heat transports and consider the SST response to a steady sun. After a time long enough for the entire ocean to reach a steady temperature, Q will vanish. The SST will then be determined purely by vertical atmospheric processes (radiative-convective equilibrium); the mean tropical temperature would be on the order of 310°K. Allowing the ocean and atmosphere to have horizontal heat transports (but keeping the sun steady) changes the tropical temperature to its observed mean value of about 298°K, and there is now steady Q's into the ocean in the tropics and out in mid-latitudes. To my knowledge, no one has performed the interactive calculation showing how SST depends on these steady Q's.

Progress has been made, however, on understanding the *variability* of SST away from its mean values. Simplifying assumptions are that the SST can be identified with the temperature of a mixed layer of depth h, and that at the bottom of the mixed layer occurs a sharp transition of magnitude ΔT, below which the ocean temperature decreases smoothly. The temperature vertically integrated over this layer is:

$$\frac{\partial T}{\partial t} + \underset{\sim}{u}\cdot\nabla T = \frac{1}{\rho C_p h}\left[-Q+Q(-h)\right] \tag{5}$$

and the turbulent heat flux at the bottom of the mixed layer $Q(-h)$ is obtained by considering the heat budget across the sharp temperature jump and is $Q(-h) = -W_e\Delta T$, where W_e, the entrainment velocity

$$W_e = \frac{dh}{dt} + w$$

is the volume rate, per unit area, that non-turbulent water is crossing the interface. There will only be heat flux across the interface when $W_e > 0$.

Variations in SST are now given by:

$$\frac{\partial T'}{\partial t} + (\underset{\sim}{u}\cdot\nabla T)' + \left(\frac{1}{h} W_e\Delta T\right)' = -\frac{1}{\rho C_p}\left[\frac{Q}{h}\right]' . \tag{6}$$

In general, the flux term in (6) *will* be important (order unity) for determining SST variability. The only exceptions to this will be in those regimes where the second or third terms can dominate.

The horizontal advection term in (6) can be large in regions of large surface current variations where the SST gradients are large. This occurs mostly in equatorial or boundary regions. For example, in the central equatorial Pacific in August, SST gradients of 1.7 K per thousand Km exist. As reversal of the South Equatorial Current on the equator can produce $u' = 40$ cm/sec giving

$$\frac{\partial T'}{\partial t} = 1.6 \text{ K/month}$$

In a 100 m mixed layer, it would take a surface flux variation of $Q' = 240$ W/m^2 to produce the same temperature change. Harrison and Schopf (1983) have argued that just such a mechanism is responsible for the SST anomalies during the spectacular 1982-1983 El Nino. Q' very likely *opposes* the rise.

The entrainment term is more difficult to discuss because vertical velocities are imperfectly known in the ocean, but it seems likely that we will find large changes in entrainment only where the mean entrainment velocity is large. Wyrtky (1981) has argued that the mean vertical velocity on the equator is of order 2 m/day and $\Delta T = 2$ K. An increase of vertical velocity as big as its mean would cool a 100 m mixed layer by 1.2 K/month corresponding to a heat flux of 160 W/m^2, which is large. A still more efficient way of cooling is by increased wind stirring causing rapid increases of mixed layer depth. Increasing the mixed layer depth by 20 m in a week would correspond to an entrainment velocity of 3 m/day which under the above conditions, corresponds to a (cooling) flux of 240 W/m^2.

In mid-latitude oceans, away from boundaries, horizontal velocities are small and temperature gradients are weak. We would expect seasonal SST variability to be described by

$$\frac{\partial T'}{\partial t} + \left(\frac{1}{h} w_e \Delta T\right)' = \frac{1}{\rho C_p} \left[\frac{Q}{h}\right]' \qquad (7)$$

(where the entrainment term must be kept to allow the possibility of rapid mixed layer deepening during winter months). SST is then determined by the joint solution of (7) and the mixed layer equations (if Q' is known) and it is clear that Q' is first order important in (7): Its accuracy therefore helps determine the accuracy than can be expected of a forecast of SST.

An even simpler situation occurs in the central Indian ocean where not only are horizontal temperature gradients small, but mixed layer depths seem to be relatively constant. McPhaden (1982) has shown that the even simpler version of (7)

$$\rho C_p h \frac{\partial T'}{\partial t} = -Q'$$

gives a good prediction of SST variability. Here a constant fractional error in Q' will produce the same fractional error in T'.

V. CONCLUSION

We have discussed the geographic distribution of the surface fluxes, the measurement accuracy currently attainable, and some considerations is relating the heat flux into the ocean to various oceanographic processes. In general, it is extremely desirable to have the heat flux into the ocean with the greatest possible accuracy, but the considerations of the previous section show that, especially in equatorial and boundary regions, there are internal dynamical readjustments of heat which can overwhelm the heat flux at the surface in producing storage and SST changes.

We conclude by emphasizing the importance of the surface wind field in all of these considerations. The wind field is not only crucial for determining the surface heat fluxes, but it is the driving force for those oceanic heat transport processes not directly dependent on the surface heat flux. A global surface wind field of known accuracy would be a major boon to oceanography.

Acknowledgments: This work was supported by NASA Grant NGL 22-007-228. I am grateful to Prof. D.E. Harrison of MIT for many wide-ranging discussions on the subject of surface heat fluxes.

VI. REFERENCES

Budyko, M.I. *Climate and Life*, Academic Press, 1974, pp. 508.

Esbensen, S.K. and Kushnir, Y. *The Heat Budget of the Global Ocean: An Atlas Based on Estimates from Surface Marine Observations*, Climatic Research Institute Report No. 29, Oregon State Univ., Dec. 1981.

Fung, I., Harrison, D.E. and Lacis, A.A. 1983. *On the Variability of Net Long Wave Radiation at the Ocean Surface*, Revs. of Geophys. and Space Phys., in press.

Gill, A.E. and Niiler, P.P. 1973, *The Theory of Seasonal Variability in the Ocean*, Deep Sea Research, 20, pp. 141-177.

Harrison, D.E. and Schopf, P.S. 1983, *Kelvin Wave Induced Advection and the Onset of SST Warming in El Niño Events* J. Phys. Oceanogr., in press.

Lamb, P.J. and Bunker, A.F. 1982, *The Annual March of the Heat Budget of the North and Tropical Atlantic Oceans*, J. Phys. Ocean., 12, pp. 1388-1410.

Large, W.G. and Pond, S. 1982, *Sensible and Latent Heat Flux Measurements over the Ocean*, J. Phys. Oceanogr., 12, pp. 464-482.

Merle, J. 1980, *Seasonal Heat Budget in the Equatorial Atlantic Ocean*, J. Phys. Oceanogr. 10, pp. 467-469.

McPhaden, M.J. 1982, *Variability in the Central Equatorial Indian Ocean. II. Oceanic Heat and Turbulent Energy Balance*, J. Mar. Res., 40, pp. 403-419.

Simson, J.J. and Paulson, C.A. 1979, *Mid-ocean Observations of Atmospheric Radiation*, Quart. J. Roy. Met. Soc., 105, pp. 487-502.

Vonder Harr, T.H. and Oort, A.H. 1973, *New Estimate of Annual Poleward Energy Transport by Northern Hemisphere Oceans*, J. Phys. Oceanogr., 3, pp. 169-172.

Wyrtki, K. 1981, *An Estimate of Equatorial Upwelling in the Pacific*, J. Phys. Oceanogr., 11, pp. 1205-1214.

EARTH RADIATION BUDGET OBSERVATIONS, OLD AND NEW

G. G. Campbell

Cooperative Institute for Research in the Atmosphere
Colorado State University, USA

ABSTRACT

A brief review of the climatology of radiation budget observations from satellites is presented, emphasizing the possibility of making estimates of energy transport. Then as an example of new data, calculations of the cloud feedback estimated from the new NIMBUS-7 ERB observations are discussed.

1. REVIEW OF OLDER OBSERVATIONS

The energy exchanged between the Earth and the Sun and space is the fundamental driving force of the atmosphere-ocean system. Measurements have been made of this energy exchange by satellites off and on for the last twenty years. Two basic measurements are made, one the emitted exitance, E, and second the reflected exitance from which the albedo, A, is derived. One can then calculate the net radiation driving the system as in equation 1.

1) $\text{Net} = (1.-A)\ I - E$

Where I = Incident solar flux
A = Albedo
E = Emitted exitance

Conveniently the reflected and emitted exitances are separate in the spectral regime so separate measurements are

C. Gautier and M. Fieux (eds.), Large-Scale Oceanographic Experiments and Satellites, 167–183.

possible for each. Generally monthly mean maps and zonal averages are discussed in analysis of the climate system.

Measurement methods

There are two measurement systems which have been used for satellite observations of the radiation budget: integrating sensor systems and the scanning systems. The integrating sensors are thermopile detectors viewing the entire earth's disk visible from the satellite. Their flat plate weighting function then allows a direct measurement of the flux at satellite altitude. These fluxes are converted into estimates of the albedo and the emitted exitance at the top of the atmosphere. This transformation depends upon assumptions about the angular properties of the emitted and reflected radiances. At best the spatial resolution of these measurements is 1000 km.

The alternative systems are scanning instruments which view the earth through telescopes of at least small apertures. These measure radiances rather than fluxes. Here again angular properties of the radiation fields are used to estimate the top of the atmosphere fluxes. These provide higher resolution, down to 1 km for some weather satellites and 100 km for the NIMBUS-7 observations presented below. These angular property assumptions are the largest error source inherent in the observations.

There are two classes of measurement budgets which have been derived, the broad band direct measurement of the fluxes and narrow spectral band measurements which require an approximate transformation to broad band fluxes (Winston et. al., 1979). We will be presenting observations primarily of the broad band 0.2- 3.8 μm and 3.8- 50. μm from Campbell and Vonder Haar (1980) and from NIMBUS-7 (Jacobowitz et al. 1983).

Error estimates

It is difficult to make a direct error estimate because there are no independent measurements of the radiation terms, so the analysis must be examined in detail. In most cases the instrument noise is small compared to the systematic errors in calibration of the instrument. An even bigger source of uncertainty is the angular models needed to convert radiance to flux or reflected flux to albedo. Finally there is a diurnal sampling bias because all the satellite systems are in sun synchronous orbits. Campbell and Vonder Haar (1980) present estimates which can be summarized as: + 8 Wm for the global averages and about + 10 Wm error in zonal averages. The larger errors are in the polar region because the terminator is often in the field of view and so the albedo is not exactly defined.

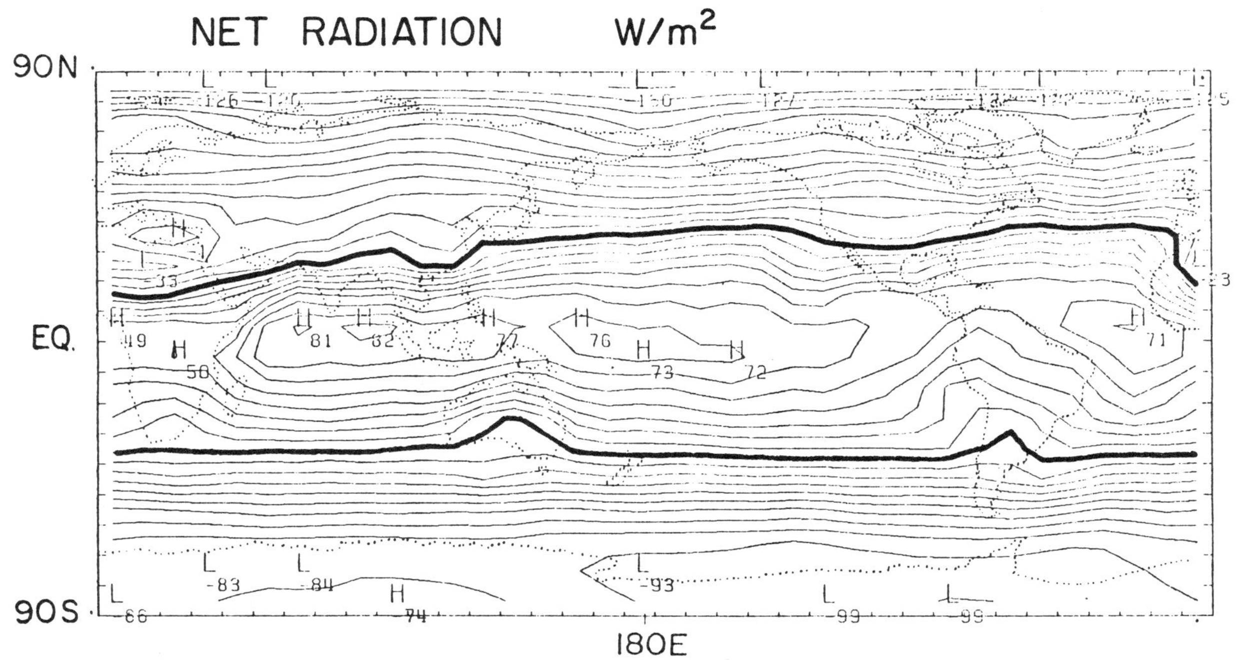

Figure 1. Annual average net radiation at the top of atmosphere from satellites. Contour Interval 10W/m^2

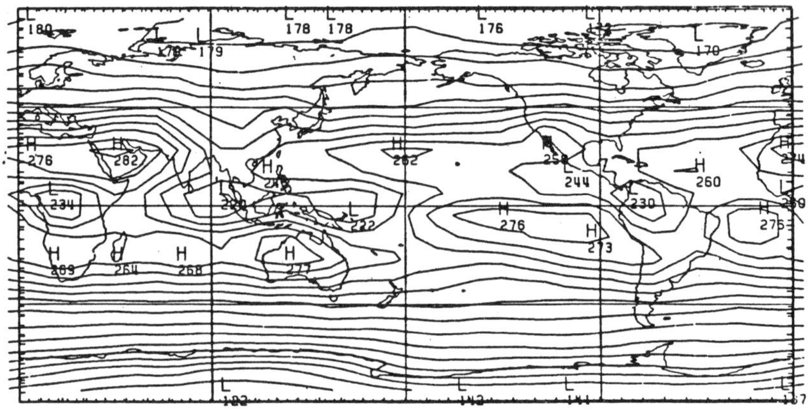

Figure 2. Annual mean emitted exitance contour plot.

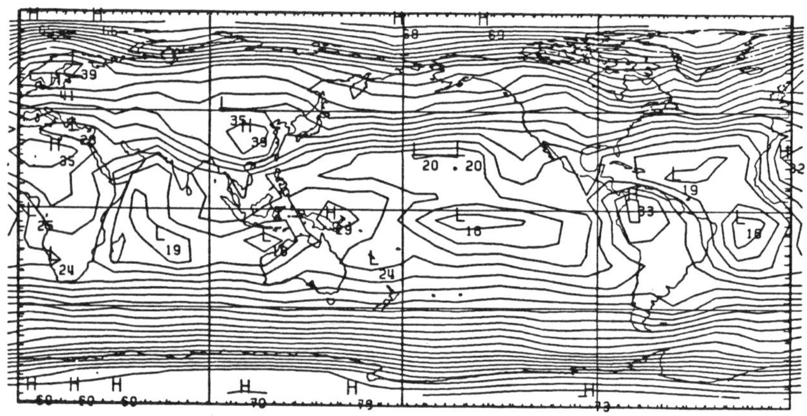

Figure 3. Annual mean albedo contour plot.

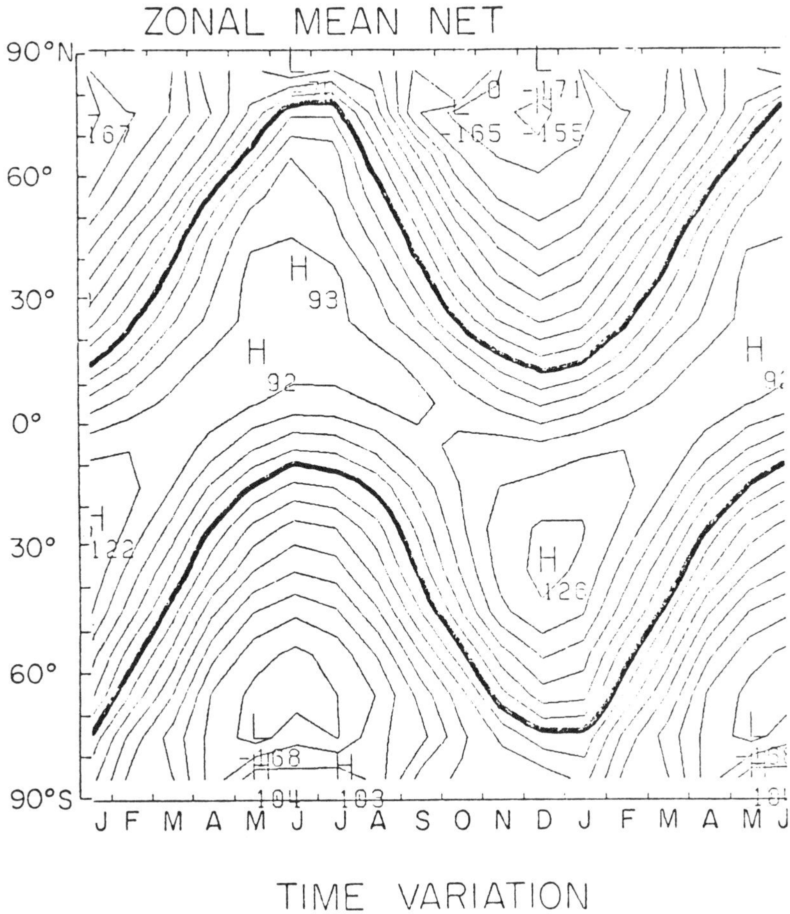

Figure 4. Contour interval 20 W/m^2. Heavy line at 0 W/m^2.

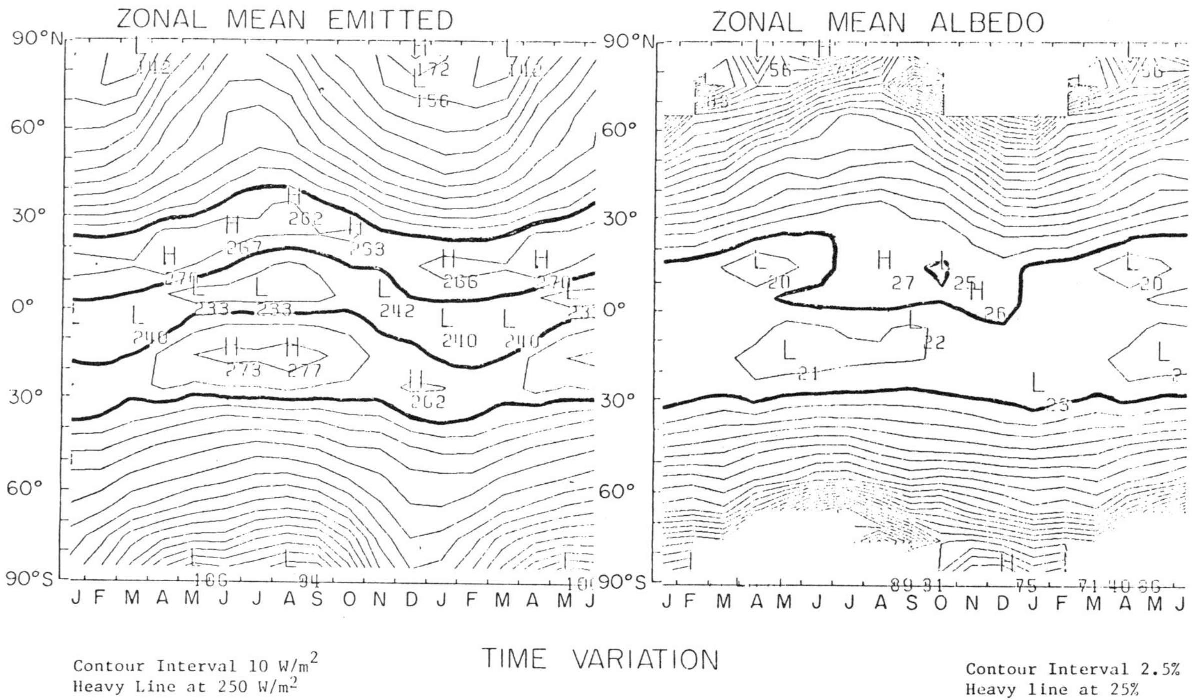

Figure 5. The time variation of the zonal means shows the seasonal change following the solar declination. 18 months are shown, 13-18 being a repetition of 1-6.

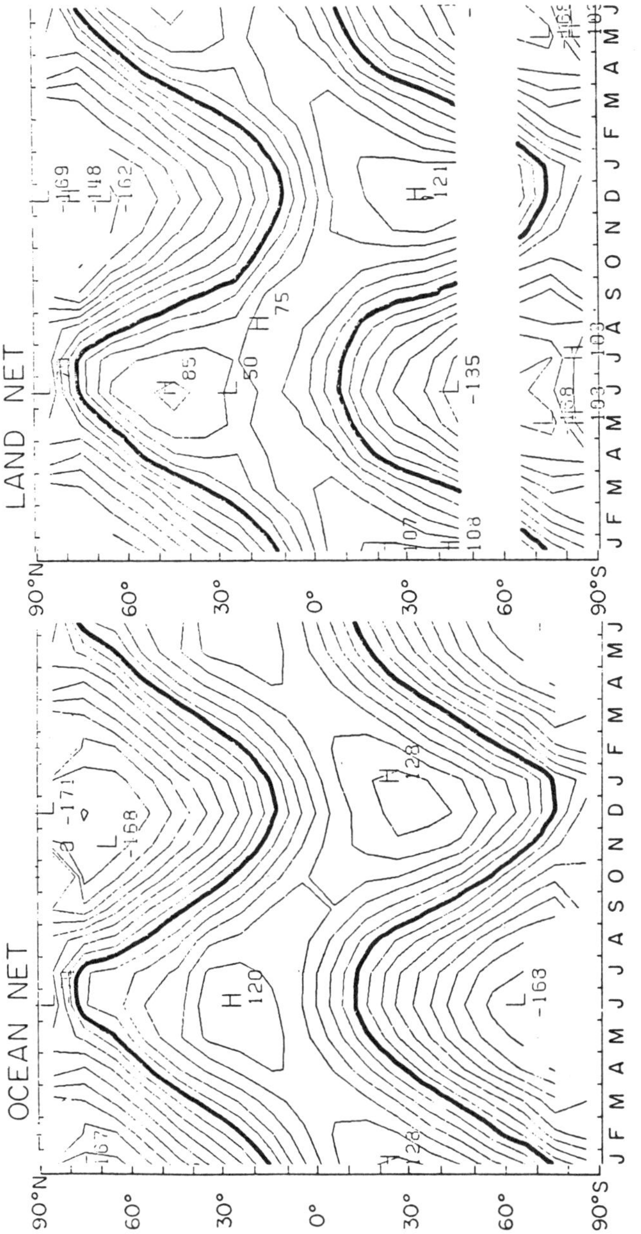

Figure 6. Separation of the annual Net radiation into land and ocean means.

2. CLIMATOLOGY OF E, A AND N.

Annual and global means

The qualitative features of the radiation budget are very familiar; solar energy is absorbed in the equatorial regions and radiation is emitted over the whole globe. This leads to a radiation surplus in the tropics and a deficit in the polar regions. Balancing the radiation process is the circulation of the atmosphere and ocean, with their transport of energy.

Figure 1 shows this more quantitatively, with the net being positive from about 30 N to 30 S. But clearly there is also a difference between land and ocean as well as between different hemispheres. Figures 2 and 3 show the two components of the budget with stronger contrast between land and ocean as well as evidence of the convergence zone with its extensive cloudiness. One can describe more details but the maps speak for themselves.

Zonal means

Zonal means illuminate the annual cycle. The net radiation (figure 4) closely follows the incident radiation, but the emitted exitance and albedo lag behind (figure 5). There is the well known lag of maximum temperature behind the maximum solar insolation; this is evident in the emitted flux. Also the albedo lags behind the solar insolation in some sense because of snow cover on the continents and because of clouds. Breaking up the net into the land and ocean components shows evidence that there is transport of energy by the atmosphere to and from the ocean regions to moderate the seasonal cycle on the continent (figure 6). This is seen in the different lag between net and the solar insolation in the two regions. Campbell (1981) goes on to calculate the residuals including atmospheric transport and energy storage and this can be shown in more detail.

Transport estimates by residual technique

These observations lead to an attempt to calculate the transports by the ocean and the atmosphere. First to balance the radiation term alone, one gets a mean annual energy transport which is almost perfectly symmetric about the equator in the annual mean. If the transport of the atmosphere (or ocean) is observed then the residual can be attributed to the ocean (or atmosphere). This was first reported by Vonder Haar and Oort (1973) with an estimate of ocean transport. Recently Oort and Peixoto (1983) have assembled observations for northern and southern hemispheres to give improved estimates of the atmospheric transports. Then they go on to calculate the residual which is assigned to the ocean transport. These are

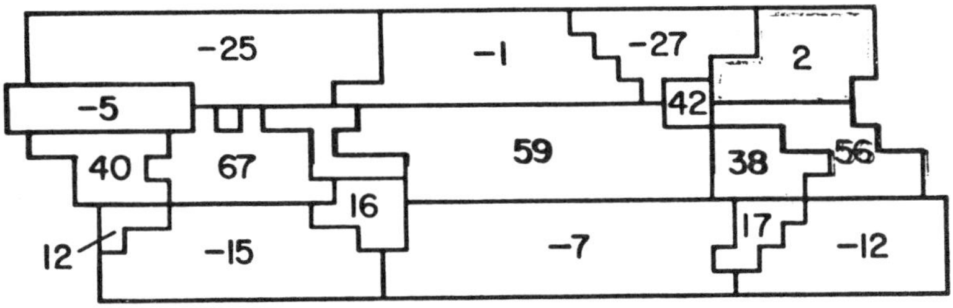

ANNUAL NET RADIATION
W/m^2

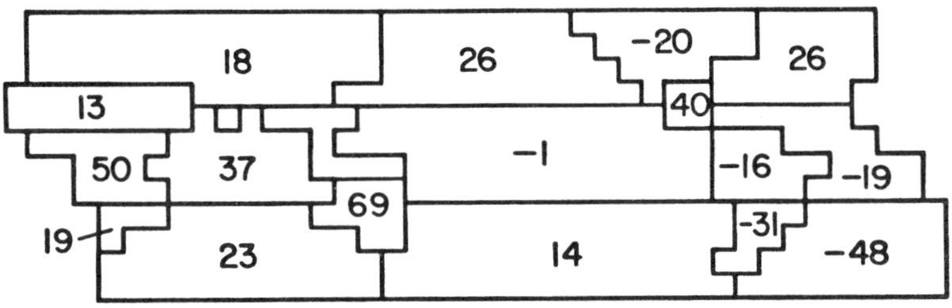

ANNUAL DIVERGENCE OF ATMOSPHERIC TRANSPORT
W/m^2

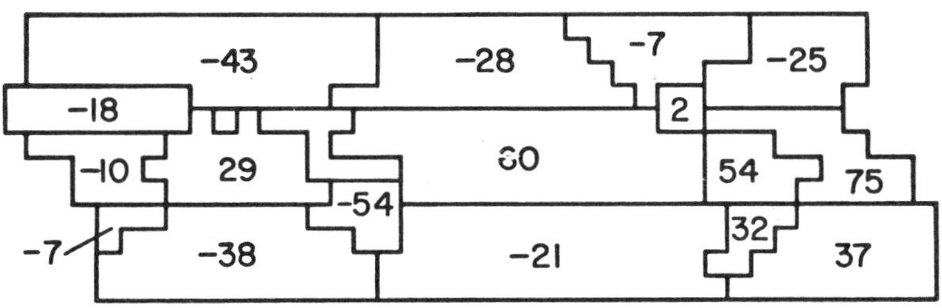

RESIDUAL = NET − div F_A
FROM BROAD BAND INSTRUMENTS
W/m^2

Figure 7. Regional budgets, Net radiation, Divergence of Atmospheric Transport and Residual. Atmospheric Term from Oort

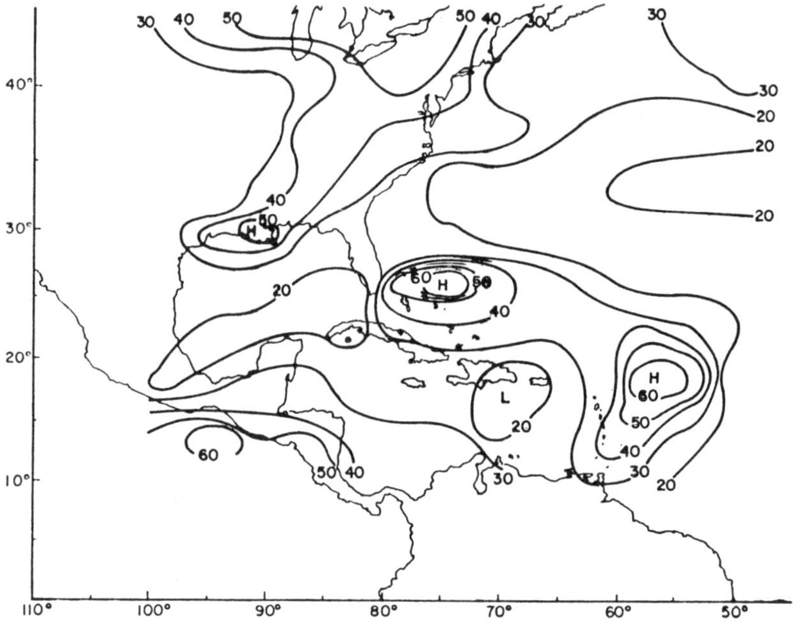

Albedo %

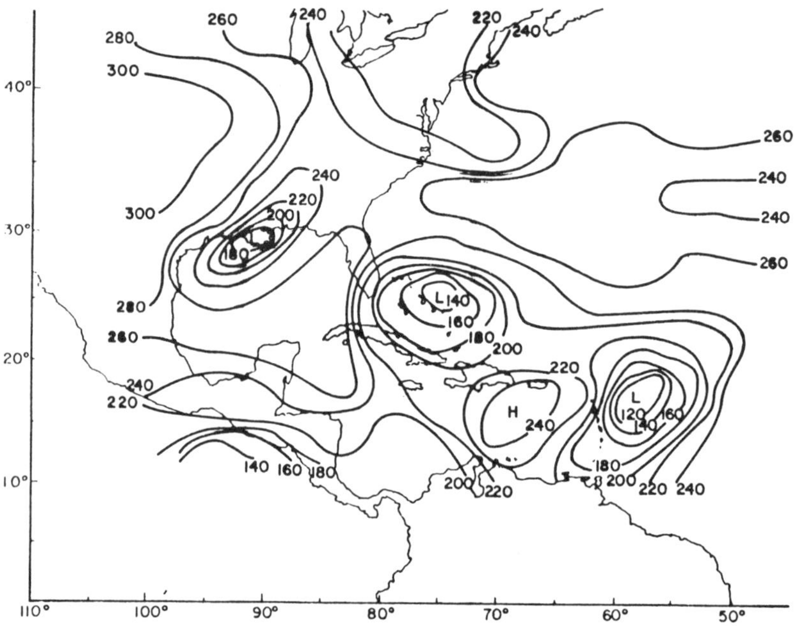

Emitted Exitance Wm^{-2}

Figure 8. Sample set of daily Nimbus 7 scanner observations. Emitted flux and albedo for September 2, 1979 from Nimbus 7.

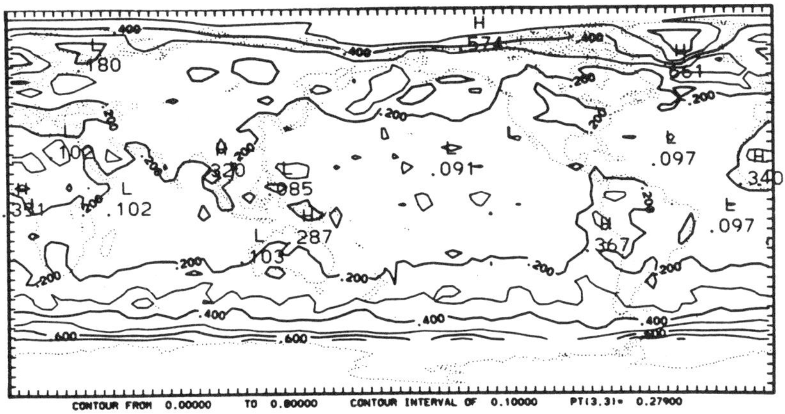

Minimum Albedo contour interval 0.1

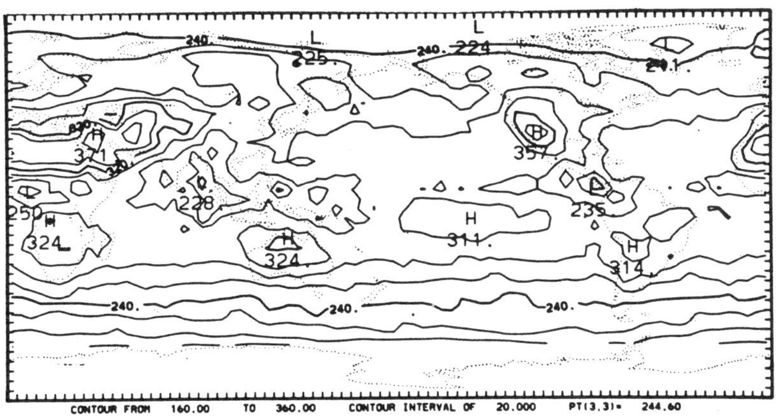

Maximum Emitted Exitance contour interval 20. Wm^{-2}

Figure 9. Minimum albedo and maximum emitted, July, 1979 from Nimbus 7.

much the same as the earlier estimates of Vonder Haar and Oort except they present an estimate of the southern ocean.

Regional budgets?

The same technique can be applied to regions to estimate something about the two dimensional energy transport. Figure 7 shows annual mean radiation budgets and mean atmospheric divergence for boxes approximating the continents. The net radiation looks reasonable so it must be concluded that the climatological atmospheric divergence of energy transport observations are not accurate enough for application of this technique. For instance, Asia should not appear as an energy source. Hopefully the data available from the Global Weather Experiment will provide enough accuracy to measure the atmospheric terms better.

3. RECENT HIGHER RESOLUTION MEASUREMENTS

Sample regional results

Observations of the radiation budget by the NIMBUS-7 scanner have recently become available (Jacobowitz et. al. 1983). These provide higher resolution, the analysis was done onto a 500 km grid. Extraction of daily data is also possible. Figure 8 from Randel (1983) shows three hurricanes in the Gulf of Mexico in September, 1979. This shows that weather events can be seen in the data. This is quite exciting because all the older monthly average data produce rather smooth fields. It might now be asked "What is the total effect of the weather and the associated cloud field on the net radiation?"

Extremes used to derive cloud effects

The primary source of high frequency time variation in the emitted flux and albedo is variations in cloud amount and type. The effects of clouds can be estimated by looking at the extremes. This has been attempted by Ellis (1976) and others. First the clear sky situation can be estimated from the minimum albedo or maximum emitted flux. Figure 9 shows the minimum albedo for July of 1979. The fact that it is moderately homogeneous in each latitude zone, except for land ocean differences implies that we have measured the mostly cloud free situation. There is good qualitative agreement with the model generated results of Breigleb and Ramanathan (1982) in their discussion of clear sky albedo. Quantitative comparison awaits future work. Similarly the maximum emitted will be representative of cloud free situation.

Let us also assume that the maximum albedo and the minimum

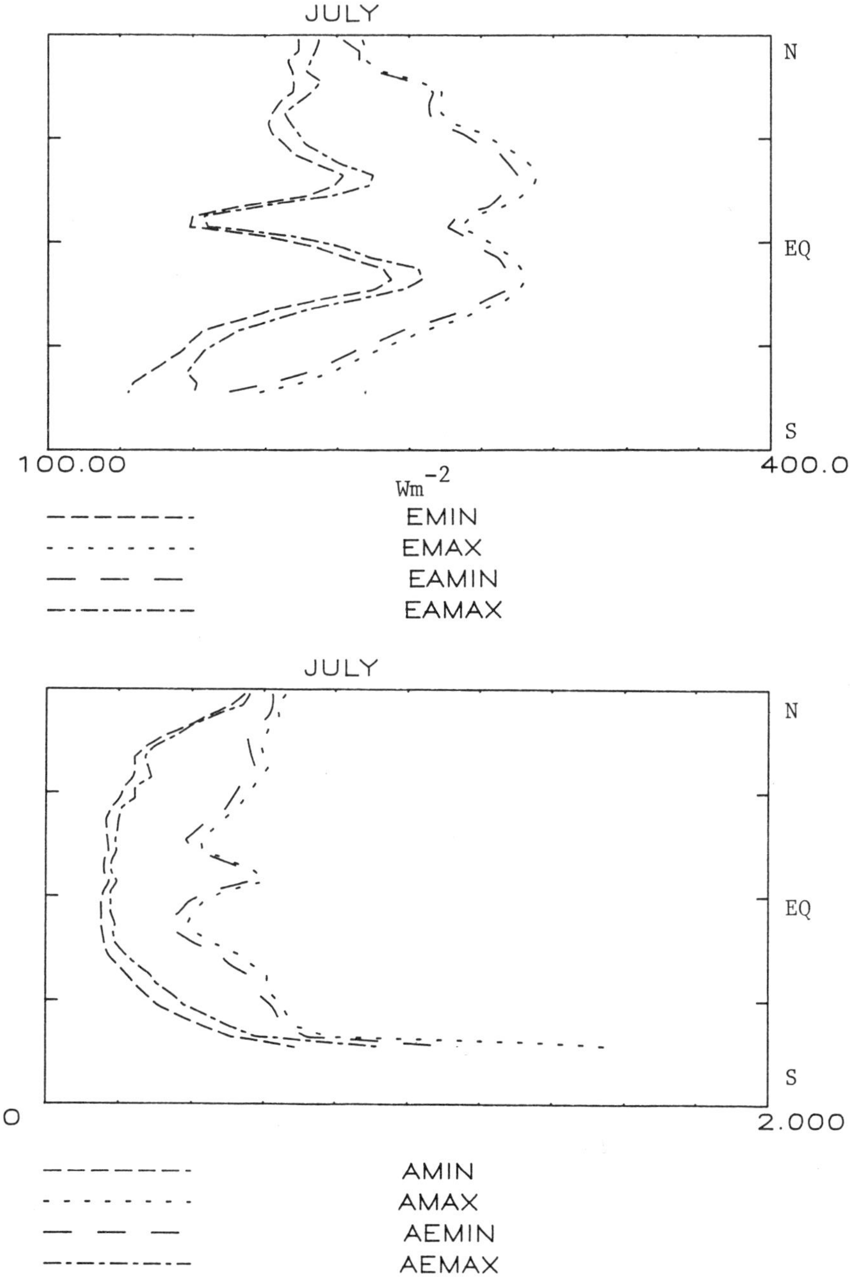

Figure 10. Albedos and emitted exitance for several extreme situations for July.

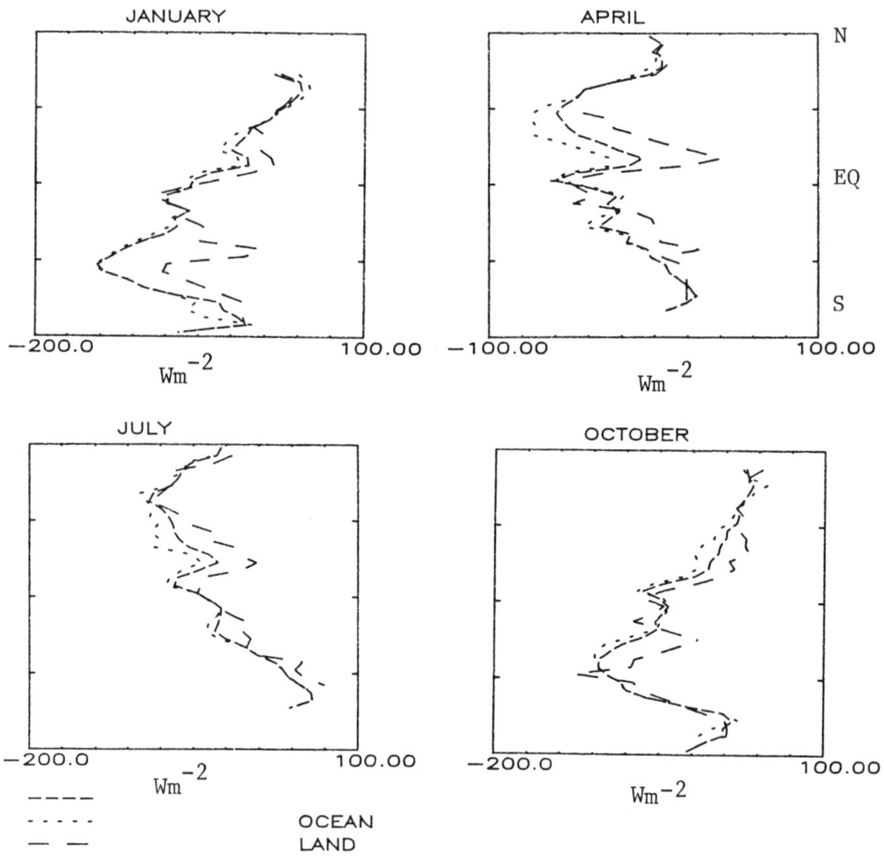

Figure 11. Zonal Average Δ net /Δ cloud for all, land and ocean.

emitted are totally cloudy. To estimate the accuracy of these assumptions we have collected the emitted flux when at the extreme of albedo and the albedo at the extreme of emission. The zonal average plots (figure 10) show comparisons. First, the albedo at maximum emitted is very near the minimum albedo, the difference being the error in the assumption that the extremes measure totally clear and totally cloudy situations.

These observations can be used to estimate the cloud feedback on the net radiation (Schneider 1972, Cess 1976). Equation 2 shows what should be measured and equation 3 shows our approximate formula:

$$2)\quad \frac{\partial\,\text{Net}}{\partial\,\text{Cloud}} = -\frac{\partial\,\text{Emitted}}{\partial\,\text{Cloud}} - \text{Incident}\,\frac{\partial\,\text{Albedo}}{\partial\,\text{Cloud}}$$

Approximately:

$$3)\quad \Delta\,\text{Net}/\Delta\,\text{Cloud} = -(\text{Emax} - \text{Emin}) - \text{Incident} * (\text{Amin} - \text{Amax})$$

Equation 3 includes the assumption that cloud amount change between minimum and maximum is 0 to 1. Certainly this is not exact because different cloud levels will have different effects on the emitted term. But let us go on to look at the results.

So far we have processed four months, one from each season, from the NIMBUS-7 data. Figure 11 shows the zonal mean variation of this estimate of the cloud feedback on the net radiation. First we see the expected effect that the summer hemisphere is dominated by the albedo term and is thus negative (January and July). Small positive feedback occurs poleward of 45 degrees in the winter hemisphere. The ocean regions have generally less albedo effect (more negative) than the land regions in the same latitude zone. The January and July situations are nearly symmetric about the equator, although the January southern ocean is more strongly negative than the northern ocean. The April and October situations are substantially different with October remaining negative for most regions, whereas the tropical land regions show positive feedback in April.

It is well known that the large heat capacity of the oceans moderates swings in the climate, certainly for the seasonal cycle for instance. Since the cloud feedback is more negative over the oceans, oceanic clouds also moderate changes in climate affecting the clouds. Land areas show less negative feedback and thus would be subject to larger response to external perturbations. Also some regions may have mean positive feedback and thus be subject to amplification of weather induced fluctuations in cloud amount.

TABLE 1

GLOBAL MEANS

	April	July	October	January	
Amin	.216	.198	.197	.198	
Amax	.506	.519	.542	.548	
Emin	189.	194.	183.	184.	W/m^2
Emax	265.	275.	269.	267.	W/m^2
		-30.	-35.	-45.	W/m^2

Global mean feedback from this estimate is strongly negative (table 1). This does not agree with the general results of Cess et al. (1982), who got near 0. Wm^{-2} for their study of broad band radiation data using the seasonal cycle to estimate the effect. We have not come to a firm conclusion about the global average effects, but the discussion of the space and time variations does not depend on the absolute accuracy of the method. We are now preparing the remaining months of an annual cycle to investigate these problems further.

4. CONCLUSIONS

Future measurements by Earth Radiation Budget Experiment of NASA will provide better measurements of regional radiation budgets. But the accuracy will still not be enough to answer whether the earth is heating up or cooling off as a whole. Regional estimates are possible, but then the partition between ocean storage, and transport in and out of a region is difficult to measure. The CAGE study concluded that atmospheric and ocean observations were just at the accuracy threshold of making useful regional energy budget measurements.

There is still much to be learned about the interaction between the radiation budget and the atmosphere-ocean system especially the interaction with clouds. The World Climate Research Program is now sponsoring the International Satellite Cloud Climatology Project with this as one of its major goals. We look forward to the comparison of the radiation budget to direct cloud estimates, dispensing with the need of indirect methods.

REFERENCES

Breigleb, B. and V. Ramanathan, 1982, Spectral and diurnal variations in clear sky planetary albedo, J. Appl. Meteor. 21, 1160-1171.

Campbell, G.G., 1981, "Energy Transport Within the Earth's Atmosphere Ocean System from a Climate Point of View", Ph.D. Thesis, Colorado State Univ.

Campbell, G.G. and T.H. Vonder Haar, 1980, Climatology of Radiation Budget Measurements from Satellites, Atmos. Sci. Paper No. 323, Colorado State Univ.

Cess, R.D., 1976, Climage change: An appraisal of atmospheric feedback mechanisms employing zonal climatology. J. Atmos. Sci., 33, 1831-1843.

Cess, R.D., B.P. Breigleb and M.S. Lian, 1982, Low latitude cloudiness and climate feedback: comparative estimates from satellite data: J. Atmos. Sci., 39, 53-59.

Ellis, J.S., 1977, "Cloudiness; The Planetary Radiation Budget and Climate", Ph.D. Thesis, Colorado State University.

Jacobowitz, H., H.V. Soule, H.L. Kyle, F. House and the NIMBUS-7 ERB Experiment Team, 1983, The earth radiation budget (ERB) experiment: an overview. Submitted to JGR.

Randel, D., 1983, "Space-Time Variations in the Earth Radiation Budget" M.S. Thesis, Colorado State University.

Oort, A.H. and J.P. Peixoto, 1983, Global angular momentum and energy balance requirements from observations: Advances in Geophysics, 25, 356-490.

Schneider, S.H., 1972, Cloudiness as a global climate feedback mechanism: The effects on the radiation balance and surface temperature of variations in cloudiness. J. Atmos. Sci. 29, 1413-1422.

Vonder Haar, T.H. and A.H. Oort, 1973, New estimate of the annual poleward energy transport by northern hemisphere oceans, J. Phys. Oceanogr., 3 169-172.

Winston, J., A. Gruber, T. Gray, M. Varnadore, C. Earnest and L. Manello, 1979, The earth radiation budget analysis derived from NOAA satellite data, June 1974 to February 1978. I and II, Met. Sat. Lab., Washington, D.C.

SURFACE RADIATION BUDGET

C. Gautier

California Space Institute, A-021
Scripps Institution of Oceanography
La Jolla, CA 92093

ABSTRACT

The incoming and net shortwave radiation (NSW) at the ocean surface can now be estimated from satellite data with a reasonable accuracy (5-10 w/m^2) for climate studies. The accurate computations are however computer intensive and strategies should be established (sampling, averaging?) to limit the computation burden.

The estimation of the net longwave radiation (NLW) at the surface is much further behind for two main reasons. First, it is more difficult to directly compute from satellite data and second, there are almost no data available for verification. It is important to start building a reliable statistical knowledge based on high quality surface measurements representative of a variety of conditions.

1. INTRODUCTION

Estimations of the incoming and the net shortwave radiation (NSW) at the surface appear now feasible from satellite measurements in the visible. This is an exciting prospect for the study of the surface heat budget since NSW is the dominant term of the net heat flux at the ocean surface in the tropical regions, where most of the atmosphere-ocean system acquires its heat and also since conventional measurements are almost inexistent in these regions. Satellite estimates offer the only reasonable approach to providing the global coverage needed for

C. Gautier and M. Fieux (eds.), Large-Scale Oceanographic Experiments and Satellites, 185–203.

climate studies. This new capability has implications for the WOCE and TOGA programs for which the knowledge of the large-scale surface forcing is crucial for the achievement of their objectives.

Several approaches to computing NSW have been recently investigated (Tarpley, 1979, Gautier et. al., 1980) and intercompared (Raphael, 1983). The most adapted ones to the climate problems are those which rely on geostationary satellite data, because they provide the highest accuracy (Tarpley, 1979), and on physical modeling (by opposition to statistical methods) of the radiative processes taking place within the atmosphere and clouds, because of their more global applications. Among the available approaches based on physical modeling, the trade off is between simple models using most of the satellite observations available, but simple physics, and more sophisticated radiative transfer models requiring assumptions regarding the properties of clouds which cannot be derived from the satellite observations. From the available preliminary studies, the accuracy of both types of approaches appear to be equivalent (Raphael, 1983). In this paper, we discuss as an example the method developed by Gautier et al., 1980 and later refined by Diak and Gautier, 1983. The emphasis however is on the results obtained by applying this method, with the purpose of illustrating some of the important features of the spatial variability of insolation and NSW.

Although insolation is an important parameter for a hierarchy of climate problems, it is not sufficient for ocean heat budget studies. As discussed earlier in this volume by Sarachik, three other components are needed: the net longwave radiation (NLW), the latent and the sensible heat fluxes. The problem of deriving the latent heat flux from satellite and *in-situ* observations is discussed by Liu and Taylor further in this volume. Because it is often a small term difficult to estimate, the sensible heat flux determination is not addressed in this volume.

The net longwave radiation (NLW) at the ocean surface is a variable that has been seldom measured because of experimental difficulties. Traditionally, studies of its variability have been based on bulk formulae applied to ship observations. The limitations of these formulae have recently been shown by various investigators and thoroughly discussed by Fung et al., 1983. The conclusions are that the use of these formulae to evaluate the net longwave radiation over the ocean is questionable. However, since no direct measurements are available, the estimates, based on these formulae, represent the basis of our present knowledge.

Although the NLW is highly sensitive to various atmospheric and cloud parameters difficult to estimate directly from satellite observations (Fung et al., 1983) it appears that satellite observations might contain the appropriate information to retrieve these cloud parameters. A possible approach based on satellite observations has been recently proposed, (Smith and Wolf, 1983) but no concomitant <u>in-situ</u> data were available for their verification. The physical basis of this method is outlined here together with a brief discussion of the kind of measurement program that should be undertaken to validate them.

2. INSOLATION AND NET SHORTWAVE RADIATION

Method

Gautier et al., 1980 have developed a method allowing the determination of the incoming solar radiation (insolation) and the Net Shortwave radiation (NSW) at the surface from calibrated geostationary satellite measurements in the visible*.

This method uses calibrated brightness measurements to determine the bulk radiative properties of the atmosphere (including clouds) and these derived properties are later used to compute the incoming radiation and NSW, taking into account the sun zenith angle. Some assumptions, not all discussed here, are implicit in this method such as the horizontal homogeneity of the clouds within a picture element. The limitations of this particular assumption are however overcome by using high resolution data (1-4 km) and, in the case of clouds smaller than that resolution, by parameterizing the possible leakage of radiation through the sides of the cloud. These improvements to the original method have been recently included (Diak and Gautier, 1983) and additional minor changes have also been implemented.

---

*Over oceanic regions we can assume that the averaged surface albedo is constant and equal to about .06, then:

$$NSW = SW - SW = (1-A)\ SW = .94*\ SW$$

where A is the surface albedo and therefore we can use insolation or NSW interchangibly for studies of the variability. In regions composed of both ocean and land, the surface albedo is not constant in space and then we must study both insolation and NSW.

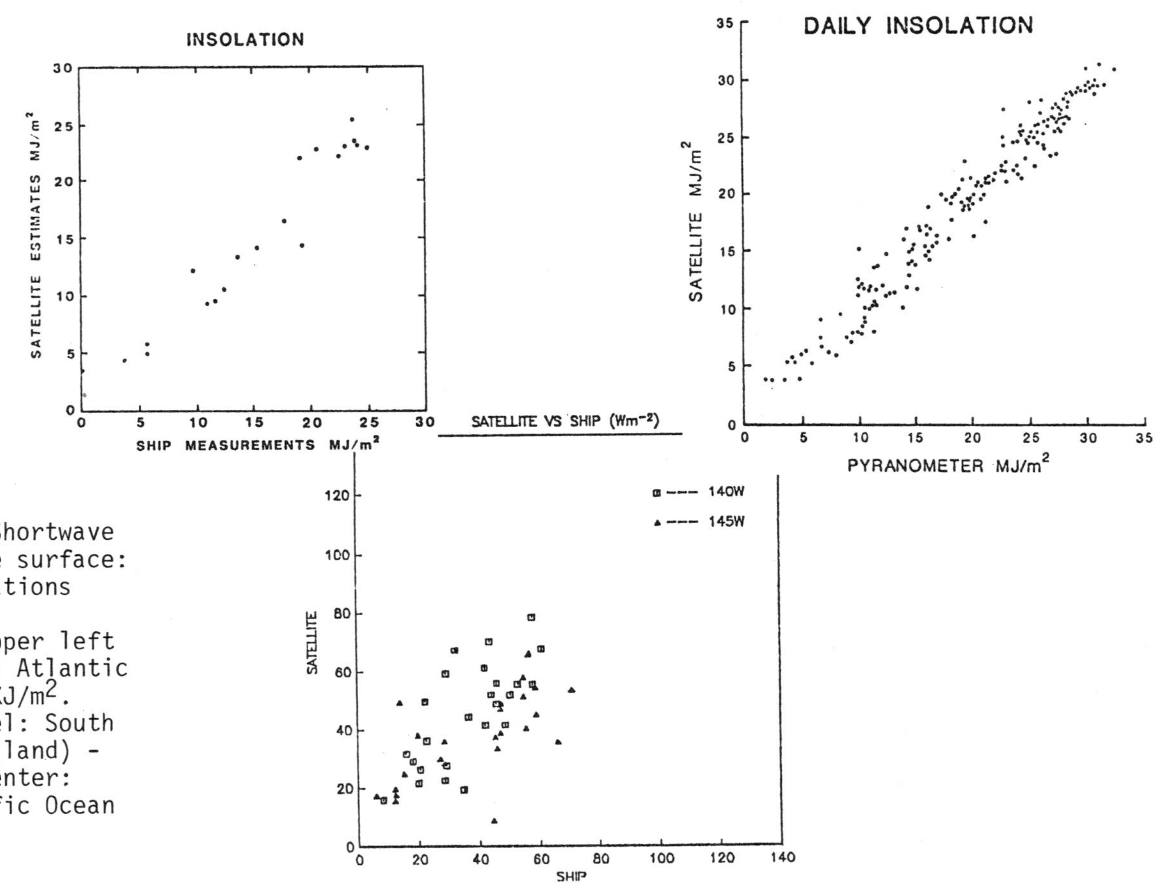

Fig. 1.
Daily Incoming Shortwave Radiation at the surface: satellite estimations versus *in-situ* measurements. Upper left pannel: Tropical Atlantic Ocean (GATE) - KJ/m^2. Upper right panel: South Eastern Canada (land) - KJ/m^2. Bottom center: North East Pacific Ocean (STREX) - W/m^2.

Comparisons with in-situ measurements

The first important result obtained by applying this method stems from the comparison with in-situ pyranometers measurements, averaged over 24 hours (daily insolation or mean daily NSW). These comparisons show an r.m.s. value for the difference between the two types of measurements of about 9% of the mean pyranometers measurements (Gautier, 1982a). This can be translated to about 17 watts/m2 in the tropical conditions of the eastern Atlantic during the GARP Atlantic Tropical Experiment, GATE (Gautier, 1982). In conditions of less insolation reaching the ground but of possibly poorer in-situ measurements, i.e. during the Storm Transfer Response Experiment, STREX, the r.m.s. of the difference is about 14 watts/m2, or about 30% of the mean surface value, (Gautier and Katsaros, 1984). A summary of these results is presented on Fig. 1, on which results from both land and ocean surfaces have been included. The second important result relates to the improvement of our knowledge of the space and time characteristics of insolation.

Error characteristics

Obviously, the interpretation of the results of Fig. 1 in the context of the larger time and space scales of TOGA and WOCE depends both on the real accuracy of the satellite estimates for individual days (for the scale over which the comparisons have been made) and on the space and time characteristics of the error fields. The estimation of the accuracy of the satellite estimates depends on the quality of the in-situ measurements used for comparison. A discussion of the quality of in-situ ship measurements during STREX is provided in Gautier and Katsaros, 1984. From the results presented above (Fig. 1) a conservative estimate of the uncertainty in the satellite derived daily insolation can be 20 w/m2 for scales of about 20 km in the tropical regions for instance. As for the characteristics of the error in the daily field, the results depend upon the existence of a bias in the satellite estimates and on the characteristics of this bias. Although a complete study of the nature of the bias remains to be made (since none of the results yet obtained have had enough coverage to uncover the existence of a regional bias), the two oceanographic comparisons performed in very different regions (Tropical Atlantic and North-Eastern Pacific) show a small positive bias of about 5 to 8 watts /m2. Based on these preliminary studies and for simplicity of the argument we will assume, at this stage, that the bias is uniform and that the error in the satellite estimates is random. Under these conditions, it is possible to quantify the accuracy improvements

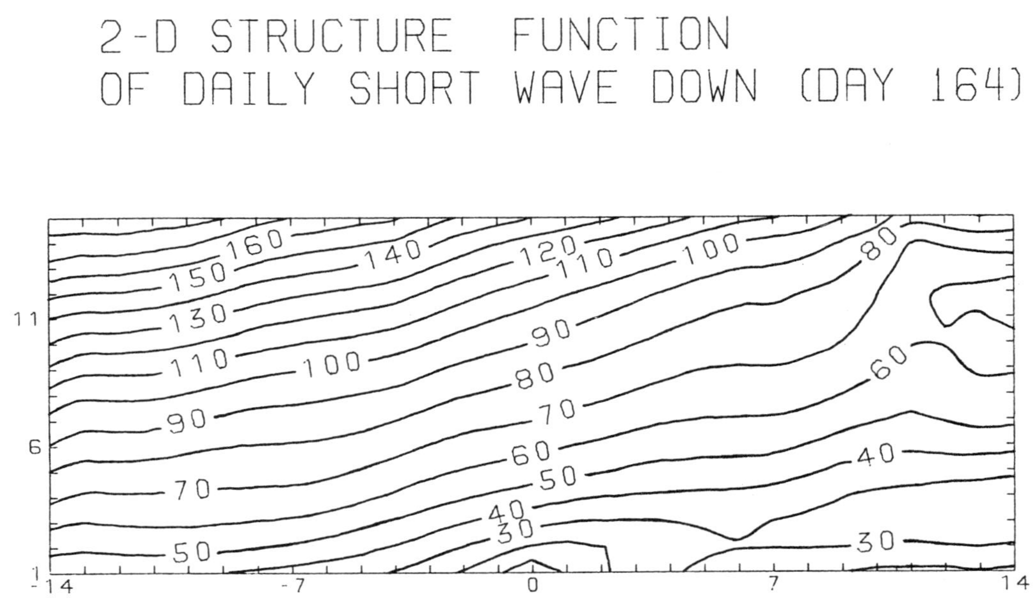

Fig. 2 Two Dimensional Structure Function of Daily Incoming Shortwave Radiation at the surface over the Indian Ocean during summer MONEX - W/m^2.

that can be brought by temporally or spatially averaging individual satellite estimates since the error correlations are equivalent to the insolation correlations.

Time characteristics

The insolation time series available over the ocean are very short, (the ocean weatherships, being the longest ones with 8 years of records) and have not yet been analyzed. However, analysis of time series in mid-latitude over land show that daily insolation estimates can, for all purposes, be assumed to be statistically independent (Gautier, 1984). The analysis of the STREX time series although relatively short, give a similar conclusion. These mean that using daily insolation estimates to compute the monthly mean insolation might allow an improvement by a factor of about 5 on the accuracy of the daily estimates.

Space characteristics

To study the spatial characteristics of insolation or NSW fields, we have used fields obtained by applying the Gautier et al., 1980 method to series of geostationary satellite data. This has been done for three different regions of the globe during different seasons. These regions and times have been selected because they corresponded to field experiments during which other types of observations were collected. Here we discuss results from the Indian Ocean region. We first computed insolation and NSW for one day and analyzed the characteristic of the field over scales of about 500 km (or squares of 15 x 15 estimates corresponding to about 32 km). The spatial variability is here illustrated by means of two-dimensional structure functions:

$$I_{DD}(\vec{d}) = \left[\overline{\left(I(\vec{x}+\vec{d}) - I(\vec{x})\right)^2}\right]^{1/2}$$

An example of such a function over the Indian Ocean during the Monsoon Experiment, (MONEX) for a randomly selected day is shown on Fig. 2. Because this function has symmetry properties, we only show half of the two-dimensional field. The spatial variability can be quite large, i.e. up to 180 w/m2 over scales of 300 km, in the cases presented here. The general tendency is for zonal orientation of the fields, as expressed by the zonal tendency of the structure function. Despite this tendency, standard averaging procedures are usually isotropic. Therefore, assuming that an isotropic spatial averaging would be performed, we investigate here the possible uncertainty reduction obtained by such an averaging procedure by studying azimuthally averaged spatial autocorrelation functions. The analysis of such functions can provide a scale of decorrelation, i.e. a scale

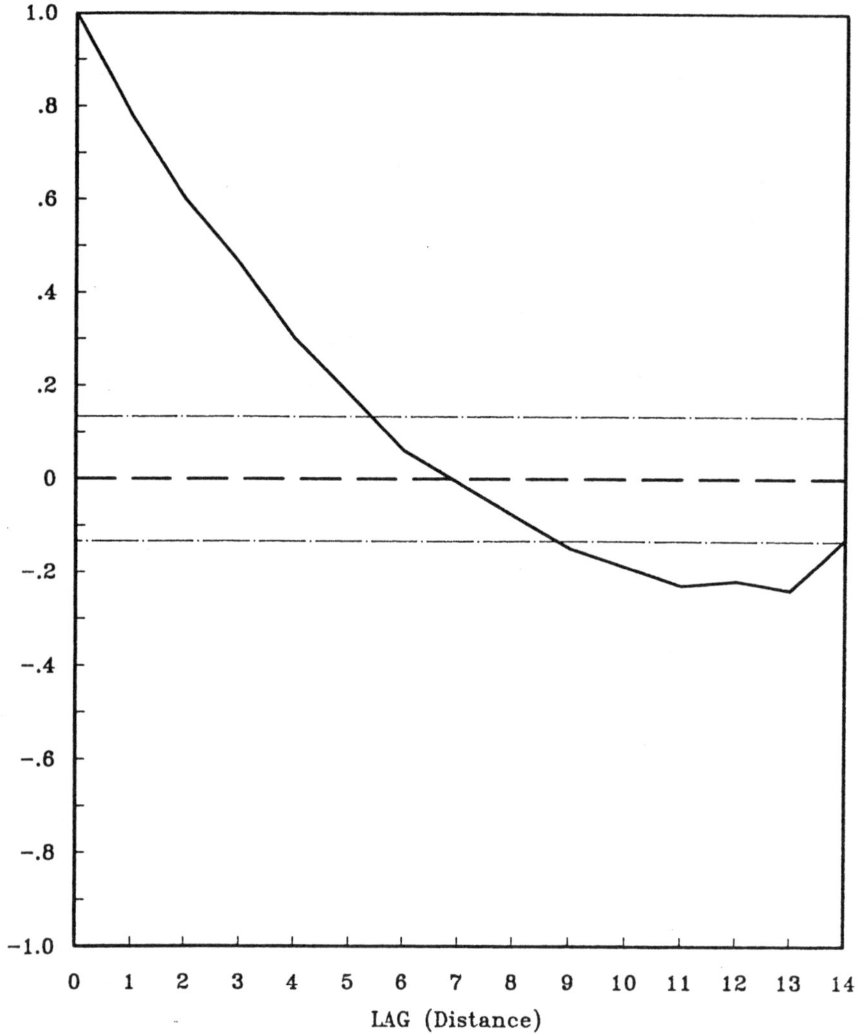

Fig. 3 Azimuthally averaged spatial correlation function distance unit 50 km

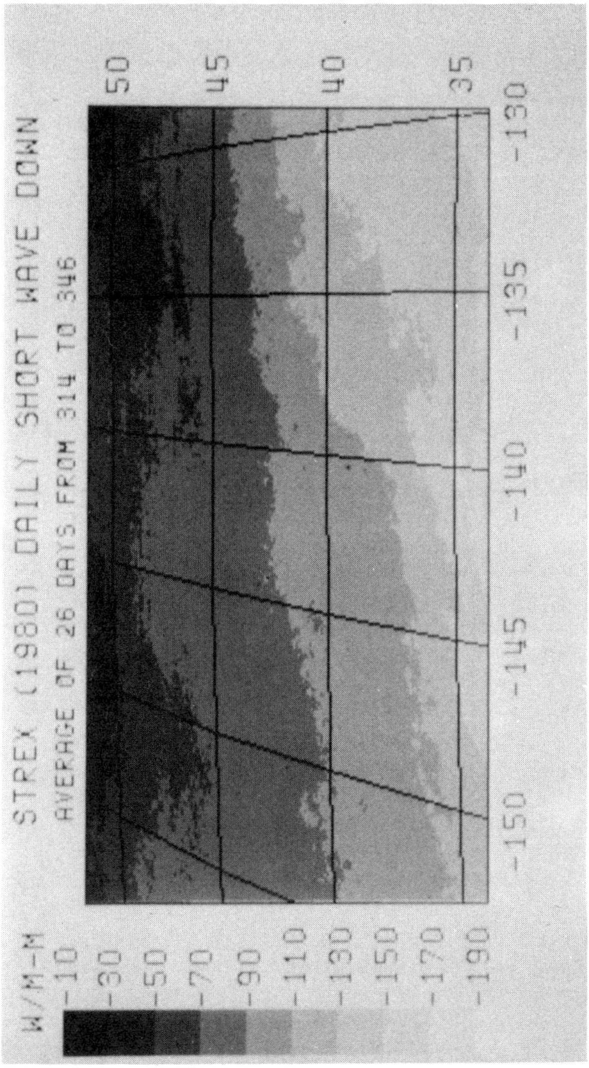

Fig. 4a Averaged short-wave radiation over the ocean (24 to 25 days) North-East Pacific Ocean (STREX) - Nov. 1980

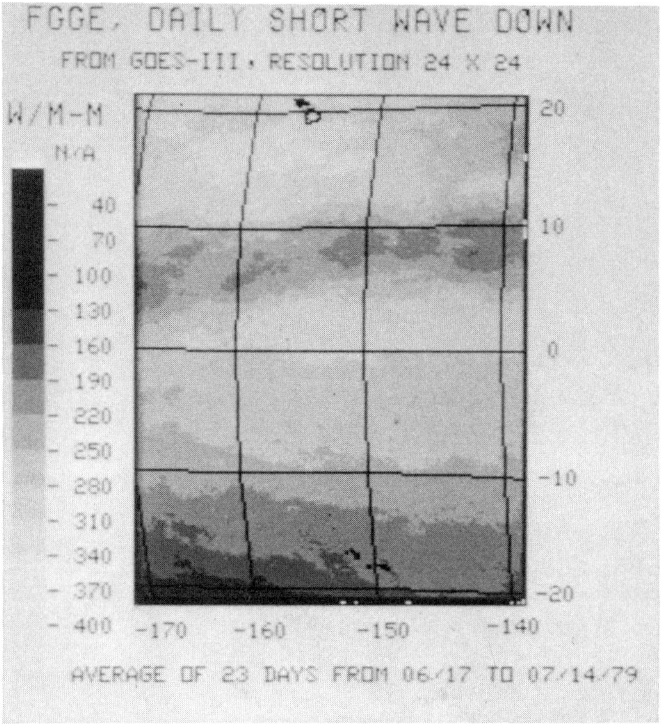

Fig. 4b Averaged short-wave radiation over the ocean (24 to 25 days)

Central Equatorial Pacific Ocean - June-July 1979

Fig. 4c Averaged short-wave radiation over the ocean (24 to 25 days)

Indian Ocean - June-July 1979

over which we can consider individual estimates to be independent. There are several approaches to estimating this scale. The approach we used was to assume a simple stochastic model for insolation variability with known lagged space correlation, and then to assume a first-order Markov process (or red-noise) for insolation, (this was first done by Leith, 1973, to determine the statistical dependence of sample values of meteorological time series). In this case the lagged space correlation is given by:

$$R\ (d) = \exp\ (-u/d)$$

where u is a parameter which can be adjusted to fit the mean correlation observed. In this case the scale of decorrelation is given by 2/u.

With the MONEX data the adjustable parameter u is found to be close to 0.3 for the averaged data of the set (see Fig. 3), which gives a decorrelation scale of about 350 km since the lagged scale of Fig. 3 corresponds to about 48 km. From these results we can anticipate an accuracy improvement of about 40% for 500 km scales, with estimates made on a 48 km scale.

Monthly insolation fields

The results of our computations for monthly averages are showed on Fig. 4a, b and c, respectively for the North-East Pacific during the Fall 1980 (STREX), the central equatorial Pacific during the summer 1979 and the Indian ocean during the summer 1979 (summer MONEX). The important aspect of these figures is the smoothness of the fields, by comparison with daily insolation fields which are much more highly variable, as was noticeable on the spatial structure function on Fig. 2. In the case of the NE Pacific, the cloud effects appear to be approximately random on the monthly time-scale and therefore an estimation of the cloud effect can possibly be obtained from continuous ship measurements at a fixed point (Gautier and Masse, 1984). In the case of the central Pacific, the most salient features are the Inter Tropical Convergence Zone (ITCZ), which reduces insolation on the average from 300 to 200 w/m2 on the monthly time scale and the South Pacific Convergence Zone (SPCZ), which even reduces insolation more drastically. If one takes typical conditions in the tropics, a horizontal gradient of heat input of 100 w^2/m over 100 km for monthly time scales could give rise to a horizontal temperature gradient of about 1° K over 100 km, if the heat were locally stored over a typical mixed layer of 50 meters (Weare, 1983). Such strong meridional SST gradient is generally not observed in the ITCZ region, therefore the heat acquired by the ocean must be constantly removed by some mechanisms such as evaporation or oceanic advection. The existence of such a gradient in association with the ITCZ is important in conditions associated with El Nino situations, during which the ITCZ and the SPCZ can be displaced by as much as several hundred kilometers. In such case the 1° K /100 km gradient in the meridional direction is not observed in the ITCZ region and therefore removal mechanisms must exist, but in a different oceanic region. Thus, if the heat is removed by oceanic advection, a strong current must exist in the abnormal ITCZ location.

The third example represents conditions prevailing at the monsoon onset and after, from June 6 to July 9 1979. The range of insolation scans approximately 150 w/m^2 (from 160 to 310) and the spatial characteristics are very similar, though with more details than those found in the atlas of Hastenrath and Lamb, 1978. The maxima and minima are found in the regions of intense monsoons, i.e., the Southwestern coast of India, South of the Equator in the ITCZ region, and in Southwest Asia. By comparison with the central Pacific, the largest gradient has a more zonal orientation around 10°N in accordance with the land-ocean contrast.

3. LONGWAVE RADIATION

Our knowledge of the longwave radiation over the ocean is very minimal and it is clear that any strategy for improving our understanding of its variability has some value and several approaches are possible. The most important steps however are to initiate surface measurements in different areas of the globe and also to start developing and testing methods which rely on satellite observations.

Physics of the problem

Fung et al., 1983's computations provide a good basis for discussing the sensitivity of NLW to the various atmospheric components. Fig. 4 and 5 have been excerpted from their paper and used here for illustration. The NLW depends on both the atmospheric profiles of temperature and humidity and on clouds, when they are present. NLW decreases exponentially with cloud optical thickness ( ) and for clouds nearly black (=4) in the infrared, NLW depends on the altitude of cloud base. Fig. 5 shows the sensitivity of NLW to cloud optical thickness as a function of cloud base altitude. For instance, a cloud of =4, with cloud base at 950 mb reduces clear sky NLW by 43 w/m2 in the mean tropical summer, by 55 w/m2 in the mean tropical winter, by 57 w/m2 in the mean mid-latitude summer, and by 69 w/m2 in the mean mid-latitude winter. A high cloud has less impact on NLW than a low cloud and the uncertainty in NLW due to cloudbase altitude is greatest for low clouds. For the same

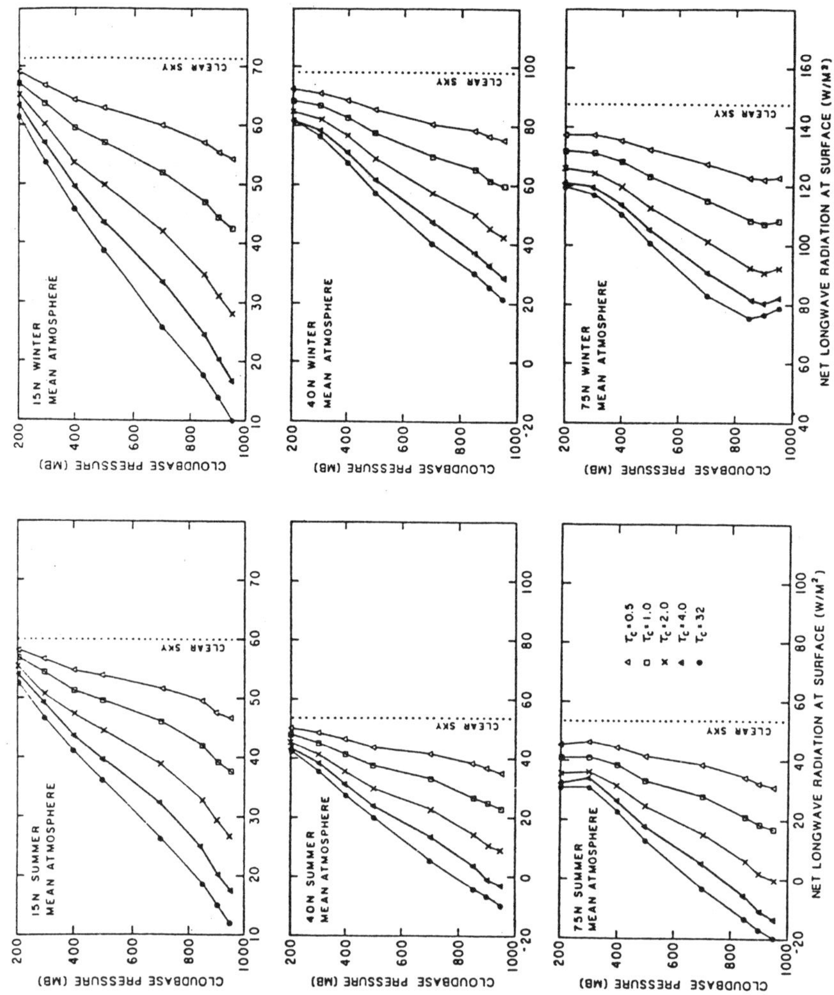

Fig. 5 Variations of NLW as a function of optical thickness (c) changes (from Fung et al., 1983)

cloud the percentage reduction of NLW depends on the environmental conditions. The results obtained by taking the same cloud with =4 and moving it up and down in an atmosphere with perturbed temperature and humidity conditions are presented on Fig. 6. These results can be interpreted in terms of the uncertainty in NLW resulting in uncertainties in the atmospheric profiles. For instance with 100% cloudiness and perturbations in temperature and humidity profiles which correspond to 2 * sigma (where sigma is the variance of the annual mean), an instantaneous measurement of NLW may vary between -14 and 114 w/m2 in mid-latitude winter. There can be large uncertainties in estimated NLW from uncertainties in the cloud estimates and in the atmospheric profiles. These results clearly illustrate the theoretical difficulties of estimating NLW from satellite observations in cloudy conditions. However, it is important to remember that our present knowledge is based on even less information about the atmosphere and clouds than those contained in the satellite observations.

A possible method

Smith and Wolf, 1983, have developed a method to statistically determine NLW from satellite observations made from the VISSR Atmospheric Sounder (VAS) on board geostationary satellites. Their approach is statistical in the sense that they use linear regression predictors to estimate NLW from actual VAS observations. These predictors are obtained from models' computations where both NLW and simulated VAS observations are computed from real atmospheric soundings. These two data sets are then linearly regressed against each other to determine the linear regression coefficients. This regression process gives a standard error estimate of the regression relations. Their statistics show that, in clear conditions, the regression explains 76.8% of the variance and has a standard error of 9.7 w/m2. In cloudy conditions the explained variance is 60.9% and the error 21.4 w/m2. Obviously these are theoretical results which must be verified against surface measurements.

Other possible methods

These results indicate that multichannel infrared instruments such as VAS (or TOVS) may contain a large part (at least 60%) of the information needed for estimating NLW. It is possible that, by adding additional constraints to the inversion process such as cloud optical thickness, it will be possible to increase the amount of variance explained in similar retrievals. Since we have seen the importance of the altitude of cloud base in the computation of NLW, an alternative approach is to try to infer cloud base altitude from the satellite observations. For

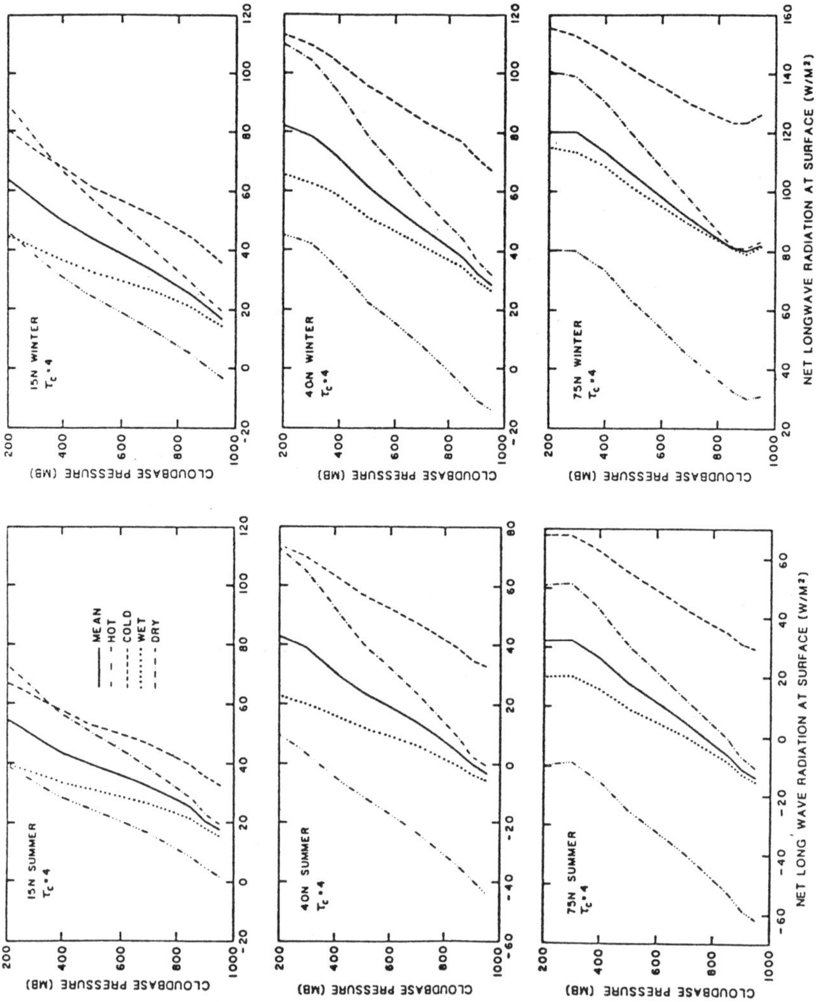

Fig. 6 Variations of NLW as a function of atmospheric temperature and humidity changes (from Fung et al., 1983).

stratiform types of clouds for example, the visible brightness is well related to the optical thickness and the liquid water content. Statistical relationships have been obtained which relate optical thickness to geometrical thickness. Thus, from the knowledge of geometrical thickness from visible observations and of cloud top altitude from infrared observations, we can infer the altitude of cloud base and compute NLW from profiles of temperature and humidity below cloud base. We will test this approach in the MILDEX experiment (October, 1983) using the atmospheric profiles measured from radiosondes and climatological profiles, and compare the results with available NLW measurements at the surface.

4. CONCLUSIONS AND RECOMMENDATIONS

Insolation and Net Shortwave Radiation (NSW) estimations from satellites are now feasible from satellite observations. Intercomparisons of methods such as those made by Raphael, 1983, are needed and high quality surface measurements for validation are required for these intercomparisons of methods. Such surface data sets should cover different regions and seasons in order to span the entire range of conditions encountered over the oceans.

Because of the large amount of data needed to be processed to obtain an adequate accuracy, programs such as the International Satellite Cloud Climatology Program (ISCCP) should consider computing insolation or NSW in their stream computations.

In the case of NLW, much research remains to be done in the satellite determination. This is starting to take place and should be encouraged. A high priority is to build reliable statistical knowledge based on high quality surface measurements representative of many types of cloud and atmospheric conditions. It is therefore strongly recommended to support surface measurements programs in connection with algorithm development research.

ACKNOWLEDGEMENTS

This work has been supported jointly by the National Science Foundation and the National Oceanic and Atmospheric Administration under grants ATM-8205817 and OCE-8214791.

I wish to express my gratitude to Serge Masse for his dedicated programming support throughout the performance of this work.

Manuscript preparation was kindly provided by Susan Hope.

Most of the computations and the image processing and display have been performed at the Scripps Satellite Oceanography Facility (SSOF).

REFERENCES

Diak, G.R. and C. Gautier, 1983, Improvements to a Physical Model for Estimating Insolation from GOES Data. J. Appl. Meteror., 22, 505-508.

Fung, I., D.E. Harrison and A.A. Lacis, 1983, On the Variability of the Net Longwave Radiation at the Ocean Surface. (Submitted to Rev. Geophys. & Space Phys.)

Gautier, C., G. Diak and S. Masse, 1980, A Simple Physical Model to Estimate Incident Solar Radiation at the Surface from GOES Satellite Data. J. Appl. Meteor., 19, 1005-1012.

Gautier, C., 1981, Daily short-wave energy budget over the ocean from geostationary satellite measurements. Oceanography from Space, 201-206. Plenum Press, New York.

Gautier, C., 1982, Mesoscale Insolation Variability Derived from Satellite Data. J. Appl. Meteor., 21, 51-58.

Gautier, C., 1982, Satellite Measurements of Insolation over the Tropical Oceans. In: Tropical Ocean - Atmosphere Newsletter, 12, 5-6.

Gautier, C., 1984, Insolation Fluctuations Statistics (in preparation).

Gautier, C., 1984, Evolution of the Net Surface Shortwave Radiation over the ocean during summer MONEX (1979): Monsoon pre-onset and onset conditions (to be submitted to Mon. Wea. Rev.).

Gautier, C. and K. Katsaros, 1984, Insolation during STREX Part I: Comparisons between surface measurements and satellite estimates (submitted for publication to J. Geophys. Res.)

Gautier, C. and S. Masse, 1984, Insolation During STREX Part II: Satellite-Derived Spatial Variability (manuscript in preparation).

Gautier, C., G. Diak and S. Masse, 1984, An Investigation of the Effects of Spatially Averaging Satellite Brightness Measurements on the Calculations of Insolation (submitted for publication to J. Appl. Meteor.).

Hastenrath, S. and P. Lamb, 1979, "Climate Atlas of The Indian Ocean Part 2, The Oceanic Heat Budget." University of Wisconsin Press, 104.

Leith, C.E., 1971, Atmospheric predictability and two-dimensional turbulence. J. Atmos. Sci., 28, 145-161.

Raphael, C., 1983, "Models for estimating solar irradiance at the Earth's surface from satellite data: an initial assessment". Canadian Climate Center, Report 83-1, 332.

Smith, W.L. and H.M. Wolf, 1983, "Geostationary Satellite Sounder (VAS) Observations of Long-Wave Radiation Flux", IGLS, Aug. 29-Sept 2.

Tarpley, J.D., 1979, Estimating incident solar radiation at the surface from geostationary satellite data, J. Appl. Meteor., 18, 1172-1181.

ESTIMATION OF LATENT HEAT FLUX WITH SEASAT-SMMR, A CASE STUDY IN N. ATLANTIC

W. T. Liu

Jet Propulsion Laboratory
Pasadena, CA. USA

ABSTRACT

The twice-daily humidity soundings from radiosonde reports in 1978 at Barbados, Puerto Rico, Bermuda and Ocean Weather Station P are used to examine the water vapor distribution in the atmosphere. The specific humidity at the lowest sounding level is found to correlate very well with the integrated water vapor in the atmospheric column at monthly time scale. An empirical relation between these two quantities is determined.

The microwave radiometer (SMMR) on ocean observing satellite Seasat provided 3 months of global observations of wind speed, sea surface temperature and integrated water vapor. With the evaluation of interfacial humidity from the sea surface temperature and near surface humidity from the integrated water vapor, the average latent heat flux can be determined by bulk parameterization method. Derivations from satellite observations are compared with those from ship reports averaged over 2° grids and the 92 days Seasat period in western N. Atlantic where the ship concentration is high and the ranges of the parameters are large.

The possibility of combining observations from different satellites is discussed and the application of this method to study interannual and spatial variations of latent heat flux in areas with large signals (tropical oceans) is suggested.

1. INTRODUCTION

The atmosphere couples with the oceans through the

C. Gautier and M. Fieux (eds.), Large-Scale Oceanographic Experiments and Satellites, 205–221.

exchanges of momentum and heat. Heat loss from the oceans changes upper ocean stratification and modify ocean circulation. Heat gain by the atmosphere from the oceans affects weather and climate. Most of the heat loss by the oceans to the atmosphere is carried by evaporation in the form of latent heat. The measurement of latent heat flux over the scales of ocean basins is fundamental in the understanding of the interannual or longer term climatic fluctuations.

There are many ways of estimating latent heat flux from research vessel or near-shore platforms: from humidity profiles (Paulson et. al., 1972), from the dissipation rate or by correlating the turbulent components (Pond et. al., 1971). Over large area and for long periods of time, the latent heat flux is often derived with bulk parameterization method from meteorological reports of merchant ships (Bunker, 1976) whose distribution is far less adequate than desired. Satellite sensors may, for the first time, provide continuous global coverage. There are various suggestions of combining satellite observations, conventional data and modeling (Chou and Atlas, 1982). This study will assess and demonstrate the potential of estimating low frequency variation of latent heat flux to a useful degree of accuracy, with available satellite observations.

The observations by the Scanning Multi-frequency Microwave Radiometer (SMMR) on the ocean observing satellite Seasat during its 3 month life span in 1978 will be used. The region of study, in western N. Atlantic (10°-70°N, 280°-320°E), was selected because it has the highest concentration of ship reports and covers a large range of latent heat flux. The ship reports (radio reports collected by the Fleet Numerical Weather Central) were used to evaluate satellite observations. Although the ship reports have questionable quality, they are the only in-situ data available that covered a large area over the Seasat period. The distribution of water vapor in the atmosphere were studies with 1 year (1978) of radiosonde data from 3 island stations: Bermuda (32°N), Puerto Rico (18°N), and Barbados (13°N), and those from Ocean Weather Station (OWS) P (50°N) in the Pacific were used to fill the data gap in high latitudes. The region of study and the positions of the 3 island stations are shown in Fig. 1.

2. BULK PARAMETERIZATION

The water vapor flux E can be determined by the bulk parameterization formula:

$$E/p = C_E U_o (Q_s - Q_o)$$

if the transfer coefficient for water vapor C_E is known and if

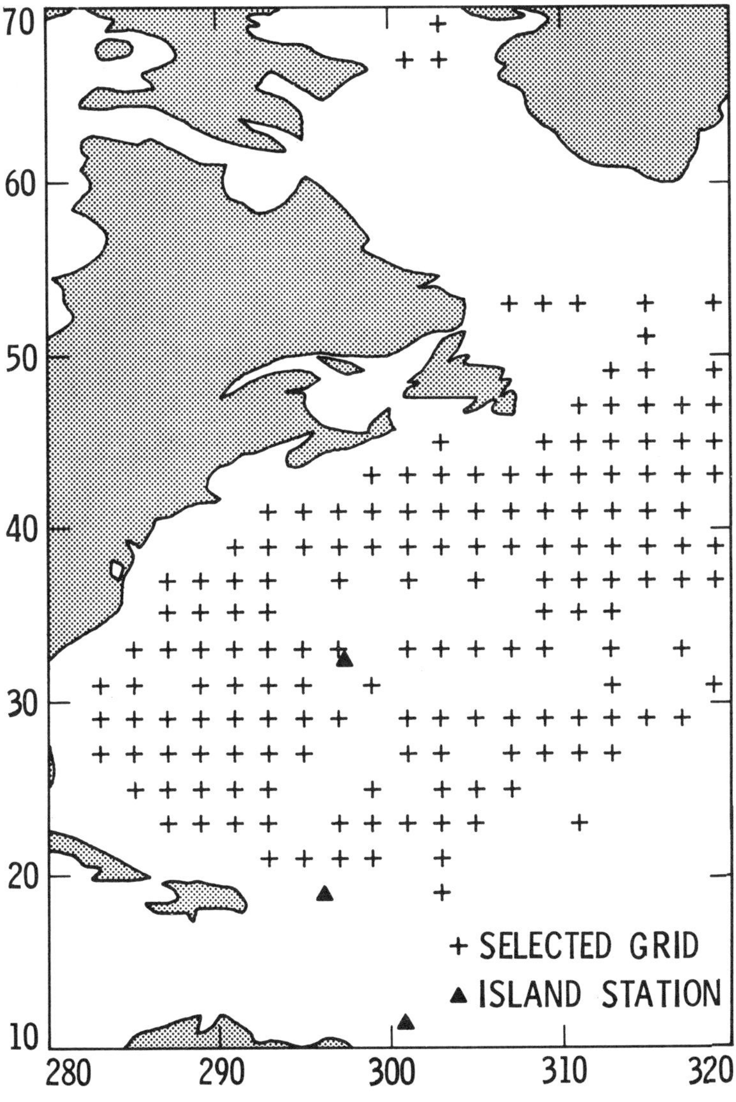

Fig. 1 Area of study with positions of the selected grids and 3 island stations.

the 3 parameters U_o, Q_s and Q_o can be measured. In the formula, p is the surface air density, Q_s is the saturation humidity at sea surface temperature, U_o and Q_o are the neutral wind speed (see Liu and Large, 1981 for definition and derivation) and specific humidity at a reference height in the atmospheric surface layer (up to 10-50m) where the flux can be assumed to be constant. The latent heat flux can be obtained by multiplying E with the latent heat of vaporization L.

The transfer coefficient C_E has been determined by regression on spot measurements (10 min to 1 hr time averages at a fixed location) up to wind speeds of about 15 ms^{-1} (Anderson and Smith, 1981; Large and Pond, 1982). The variation of the coefficient with reference height, atmospheric stability, different mechanisms of transport at the interface can be accounted for by using atmospheric surface layer models (Deardorff, 1968; Liu et. al., 1979), essentially by inversion of similarity relations. With limited amount of field data, the uncertainty in C_E is about 10-20% over moderate range of wind speed. The success of ocean surface heat budgeting by using bulk parameterization (McPhaden, 1982) suggests that the uncertainty in C_E may be due to random error that can be reduced in temporal and spatial averaging when only low frequency or large scale variations are required.

Esbensen and Reynolds (1981) evaluated heat fluxes by bulk parameterization method with individual set of parameters from ocean weather stations and found that the monthly averages of these fluxes do not differ significantly from average fluxes obtained by applying the same parameterization method to the time averages of the parameters. Similar comparison with spatial averages over 100Km areas were made by Liu and Katsaros (1983) using aircraft data during the Joint Air-Sea Interaction (JASIN) experiment in the North Sea and found little difference in the two averaging methods. The results are important because only time averaging Q_o will be used in this study as discussed in the following sections. Satellite observations are generally spatial averages and observations by different satellite sensors may not occur over a certain area at the same time. In applying a combination of satellite parameters to determine fluxes, it is essential that the bulk parameterization method which is developed from spot measurements be applicable to time and spatial averages.

Seasat-SMMR was designed to measure U_o, sea surface temperature T_s, and the integrated water vapor in the atmospheric column W (Lipes, 1982). These 3 quantities can be measured by a number of satellite sensors, but, at present, no satellite sensor has the vertical resolution to provide humidity at a level in the atmospheric surface layer or even in the

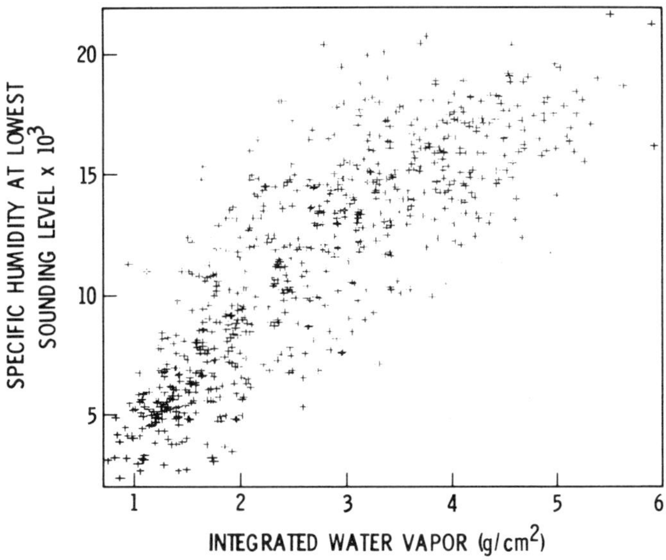

Fig. 2 The specific humidity at the lowest sounding level versus the integrated water vapor per unit area as obtained from the twice-daily radiosonde reports at Bermuda in 1978.

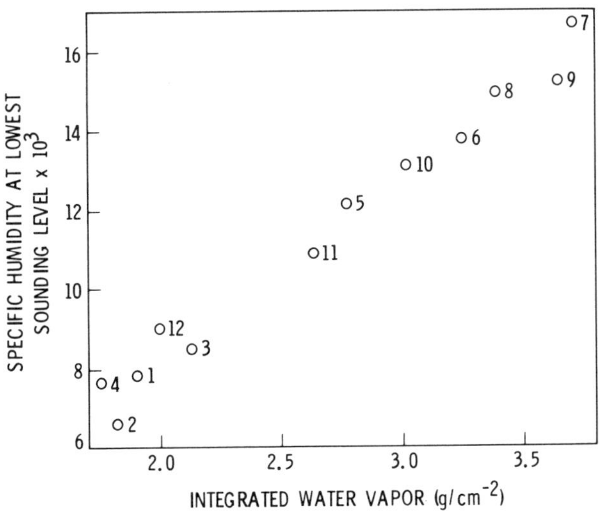

Fig. 3 As in Fig. 2 except only for monthly averages.

planetary boundary layer (lowest kilometer) where conventional bulk parameterization methods are valid (Liu, 1981). The integrated water vapor as observed by satellite sensors has to be related to the near surface humidity for the bulk parameterization method to work.

3. RELATING THE INTEGRATED WATER VAPOR TO NEAR SURFACE HUMIDITY

In order to relate W to Q_o, the twice-daily radiosonde reports from 4 stations were examined. Each humidity sounding was integrated to obtain W. The specific humidity at the lowest level was used for Q_o. The data of Bermuda, which has the largest range of water vapor, are shown in Fig. 2. There is obvious correlation but the scatter is large. The correlation increases dramatically for monthly averages as shown in Fig. 3. To show the annual variation of water vapor, the 12 months of the year are represented by the number beside the data. The monthly-averaged relative humidity at the lowest sounding level changes from 57% to 77%, indicating that the annual variation in humidity is not entirely due to variation of air temperature. The data from the other 3 stations show the same results: the monthly averages show good correlation between Q_o and W. The reduction in the r.m.s. deviations from the linear regression (the scatter) is demonstrated in Table 1. The scatter of the monthly averages for OWS P, Puerto Rico, and Barbados is approximately 1/3 of those of twice-daily data. There is even more reduction for Bermuda. For p=1.23 Kg/m^3, L=2.46 x 10^6 J/Kg, $C_E=1.2x10^{-3}$ and a typical value of U_o=7 m/s, an error of 1 g/Kg in Q_o is equivalent to an error of 25 w/m^2 in latent heat flux.

Table 1. The r.m.s. deviation (in g/Kg) from the Q versus W linear regression for radiosonde soundings at 4 stations.

STATION	TWICE DAILY	MONTHLY AVE
OWS P	0.93	0.29
BERMUDA	2.52	0.48
PUERTO RICO	1.79	0.63
BARBADOS	1.05	0.38

The monthly averages for the 4 stations are plotted together in Fig. 4. The figure shows the variations of atmospheric water vapor with latitude in addition to the annual variation at each station. The annual variation at Bermuda is only slightly less than the latitudinal variation of the 4 stations.

The gradient of the regression appears to be largest at mid-latitude. Considering the range of the data, the scatter is very small (0.78 g/Kg) and the correlation is 0.99. The result

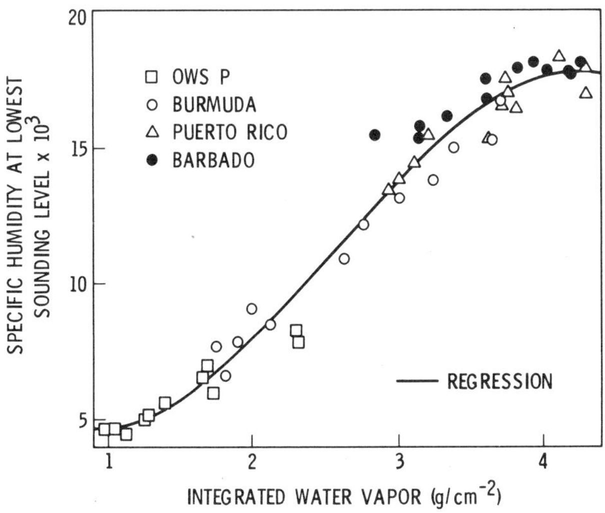

Fig. 4 As in Fig. 3 except for 4 stations instead of just for Bermuda.

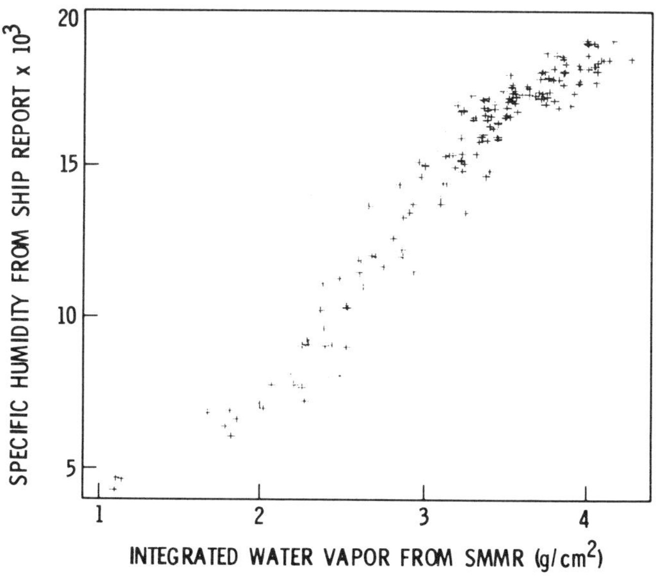

Fig. 5 The specific humidity as determined from the dew point temperature of ship reports versus the integrated water vapor per unit area as measured by the SEASAT-SMMR. Each data is an average over 2° areas and the 92 days period.

is even more remarkable when it is compared with Fig. 5. In Fig. 5, the horizontal axis is the integrated water vapor measured by Seasat-SMMR and the vertical axis is the near surface humidity derived from the dew point temperature of ship reports. Each data point is an average over 2° latitude by 2° longitude areas and the 92 days period (July 11, 1978 - October 10, 1978). The grids selected for comparison are shown in Fig. 1. Those adjacent to land were not included to avoid land contamination of the SMMR data. Each grid selected has more than 20 ship reports on dew point temperature after selection by a quality control procedure (discussed later). The data in Fig. 4 and 5 were measured by different methods and averaged over different spatial and time scales but they agree very well. The agreement demonstrates not only that near surface humidity can be derived from satellite observation but that the SMMR is a good sensor for atmospheric water vapor (at least for monthly averages) as found out by Katsaros et. al. (1981).

The radiosonde soundings demonstrated that there is good correlation between monthly-averaged near-surface humidity and integrated water vapor. In order to shed some light on why monthly averaging works and whether good correlation exists at shorter time scale, the radiosonde soundings at Bermuda were examined for the variation of water vapor distribution in the atmosphere.

4. VARIATION OF WATER VAPOR DISTRIBUTION IN THE ATMOSPHERE

The average of all the humidity sounding interpolated to standard level at Bermuda for 1978 is shown in Fig. 6. It decreases smoothly and monotonously with increasing height following closely a power function form first suggested by Smith (1966):

$$Q/Q_o = (P/P_o)^{\lambda}$$

where Q is the specific humidity at pressure level P; Q_o and P_o are the specific humidity and pressure near the surface; is constant over a latitude belt. This power function is equivalent to a relation between Q_o and W:

$$Q_o = CW$$

Where $C=g(\lambda+1)/P_o$ and g is the acceleration due to gravity. Individual sounding often does not decrease smoothly with height but sometimes exhibit a kink at the top of the boundary layer and undergoes diurnal and synoptic variations. The variation is reflected in the kinks of the vertical distribution of variance shown in Fig. 6, the lower of which may represent variations at the top of the boundary layer.

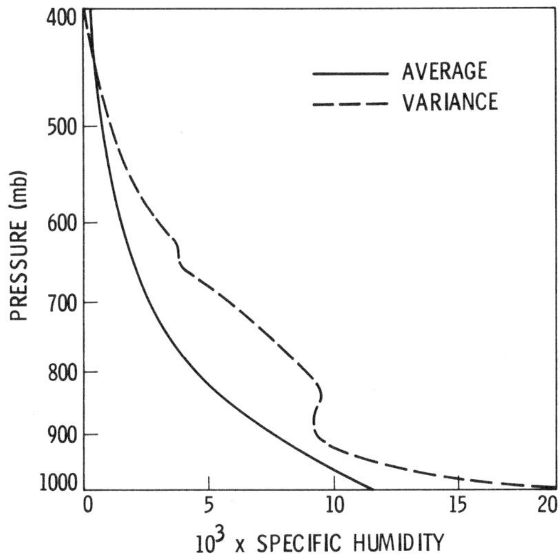

Fig. 6 Vertical profiles of the average and variance of humidity soundings of Bermuda in 1978.

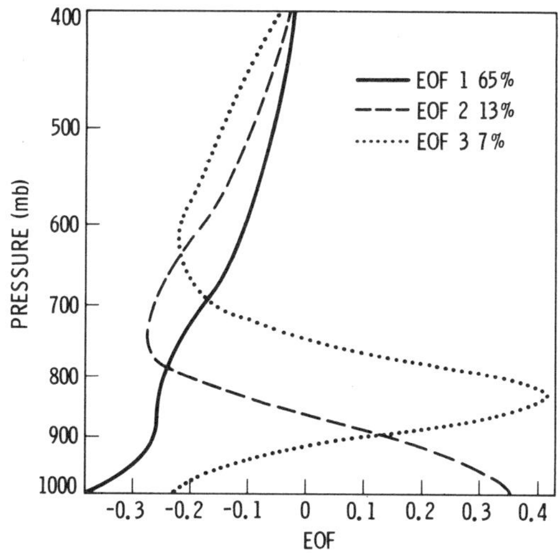

Fig. 7 Three empirical orthogonal functions accounting for most variance of the humidity soundings at Bermuda in 1978.

The empirical orthogonal functions (EOF) are the most efficient way of representing uncorrelated modes of the variance (Lorenz, 1977). The first 3 EOFs, accounting for 65%, 13% and 7% of the total variance respectively are shown in Fig. 7. The first EOF decreases smoothly with increasing height except for a small bump at 800 mb and the others have peaks above the surface. Fig. 8 shows the band-averaged spectra of the time-dependent amplitudes of the EOF plotted in a variance conserved form. EOF 1 has a spectral gap at time scale of about 30 days and the other EOFs have little variance at time scales longer than a month. When the EOF analysis was performed on the 30 days running mean, almost all the variance could be accounted for in the first EOF and the variance at 800 mb was reduced as shown in Fig. 9.

The analysis demonstrates that much of the variance of humidity above the surface has time scales of less than a month. Averaging reduces these variances and increases the correlation between Q_o and W. At time scale shorter than a month, the error in predicting Q_o from W increases. One way to increase the accuracy of prediction for time shorter than a month is to introduce a second predictor, in addition to W, which can account for the variance at higher level (e.g. boundary layer height).

5. COMBINING THE SATELLITE OBSERVATIONS

With an empirical relation between Q_o and W determined from radiosonde soundings as shown in Fig. 4, average U_o, T_s, and Q_o can be obtained from SMMR observations and averaged for each 2° areas and the 92 days period. These 3 parameters were compared with those derived from ship reports in the grids which have more than 20 reports of the respective parameters. Some of the pertinent results are summarized in Table 2, where N is the number of grid averages, BIAS and RMS are the mean and rms difference between satellite and ship data, and STD is the rms difference between ship data and the satellite data with the BIAS removed. The formulation of Lowe (1976) was used to determine Q_s from T_s. From the Clausius-Clapeyron equation (Fleagle and Businger, 1963), an error of 1^oC in sea surface temperature can be translated to about 6% error in Q_s at 25^oC. The horizontal axis of the Fig. 10 represents a combination of the parameters derived from SMMR observations. The vertical axis of Fig. 10 represents the latent heat flux derived from ship data with the bulk parameterization method of Liu et. al. (1979). Each point in the figure is an average over a 2° grid and the 92 days period. Only those grid with more than 20 ship reports on all 3 parameters are used. The positions of the selected grid are the same as those shown in Fig. 1. The scatter about the linear regression is 34.9 w/m^2.

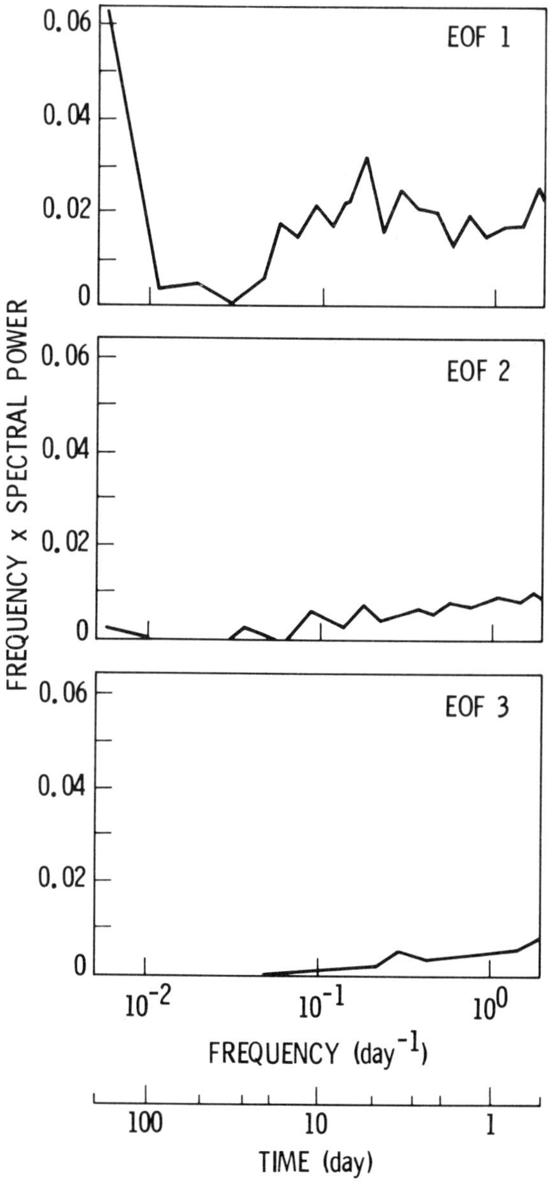

Fig. 8 Band-averaged spectra of the time-dependent amplitudes of the 3 EOF shown in Fig. 7.

Table 2. Comparing average satellite and ship data.

PARAMETER	N	BIA	RMS	STD
U_o	212	1.52	2.02	1.33
T_s	203	0.12	1.19	1.18
Q_o	161	-0.48	0.97	0.84

A selection procedure was designed to control the quality of ship reports. The first step was an attempt to remove reports with position and time errors. Data with positions on land were first discarded. The displacement and time elapsed between consecutive reports by the same ship were used to evaluate mean ship speed. When the speed exceeded 35 knots, the second report was discarded. The second step removed data outside reasonable ranges. The ranges set for wind speeds was 0 to 90 knots, for sea surface temperature, it was 0-40°C. Those for air temperature and dew point temperature were -10 to 40°C. In the last step, the mean and standard deviation for each 2° grid and the 92-days period were calculated. Criteria of 3.6 and 2.0 standard deviation from the mean were chosen for wind speed and temperature respectively. Data outside these ranges were discarded.

The SMMR winds are higher than ship winds and the difference increases with increasing wind speeds. Merchant ships, however, tend to avoid storms and the quality control procedure discards extreme data. Thus, ship winds used in this study may be biased low. The difference in sea surface temperature decreases with increasing sea surface temperature and above 15°C, the sea surface temperature measured by SMMR and from ships agree very well. When a temperature-dependent bias is removed from SMMR T_s and wind speed dependent bias is removed from SMMR U_o, the r.m.s. difference between satellite and ship data are 1.02 m/s in U_o and 0.83 in T_s. When, for the abscissa in Fig. 10, U_o is replaced by detrended U_o and Q_s is replaced by the derived from the detrended T_s, the scatter in latent heat flux is reduced to 30 w/m^2.

6. POTENTIAL APPLICATIONS

Satellite observations have characteristics quite different from those of <u>in-situ</u> measurements. But, before a method more attuned to space borne sensors is developed, the well-established bulk parameterization method appears to be most promising in leading to oceanic and climatic applications.

Estimating ocean surface parameters from space is still in very early stages of development and the accuracies, in many cases, are still crude. Satellite observations cannot be effectively applied to all areas with scientific impacts but

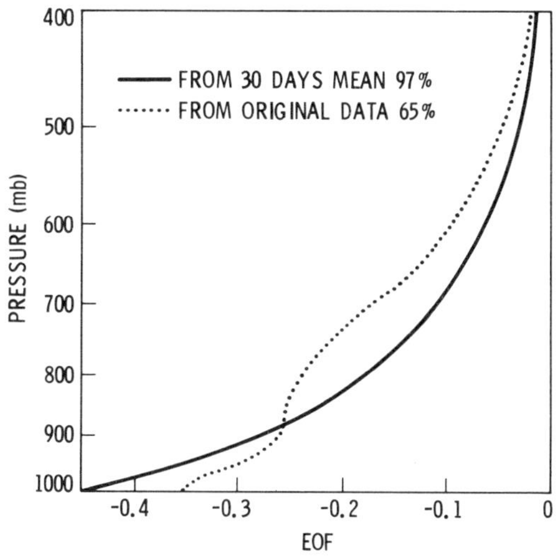

Fig. 9 Comparison of EOF 1 of the original humidity soundings and those of the 30 days means.

should first be tried in areas with large signals.

Significant relationships have been found between the interannual variations in Tropical Pacific and global atmosphere (WCRP, 1983). Latent heat exchanges between the ocean and the atmosphere is believed to be an important physical link. Niiler (1982) found that the difference in heat flux between the December of a normal year (1973) and the December of an anomalous year (1972) can be as large as 200 w/m^2. Fig. 11 shows the temporal and spatial variation of latent heat flux at equatorial Pacific as derived from the 20 years means of Weare et. al. (1980). Fig. 11a shows the annual variation at 135°W on the equator and Fig. 11b shows the spatial variation on the equator for August. The temporal and spatial variations have about the same magnitude, 120 w/m^2. The interannual standard deviations for 20 years, in many cases, also exceeds 100 w/m^2. In tropical oceans, the mean and variation appear to have similar magnitude. An accuracy of 15 w/m^2 would be about 10% of the signal.

In tropical oceans, consider the typical values of U=7 m/s and $Q=Q_s-Q_o=6$ g/Kg which give a latent heat flux of 152 w/m^2. An accuracy of 20%, equivalent to 31 w/m^2 in latent heat, requires either 1.5 m/s in U or 1.2 g/Kg in Q. These, or even higher, accuracies are achievable particularly for a region with high quality in-situ measurements serving as references to calibrate satellite observations.

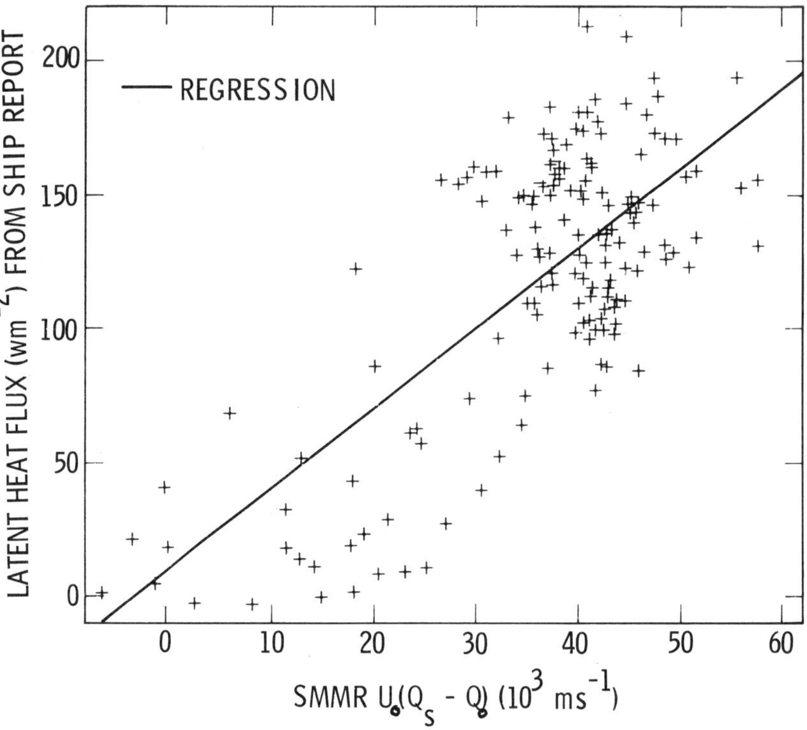

Fig. 10 Surface latent heat flux evaluated from ship reports compared with derivations from SEASAT-SMMR observations. Each point is an average of 92 days and 2° area.

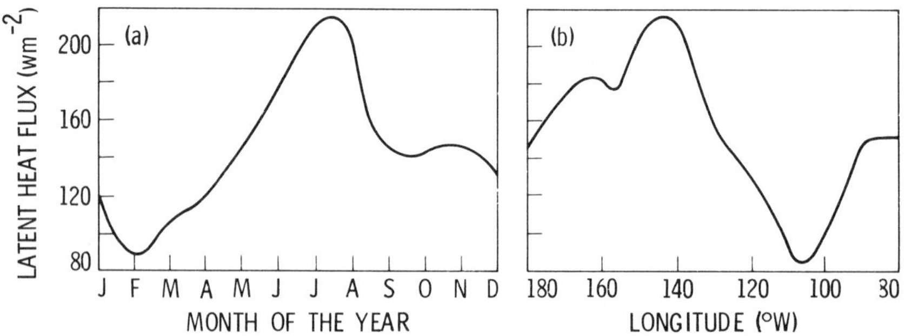

Fig. 11 Temporal (a) and spatial (b) variation of latent heat flux at equatorial Pacific, estimated from 20 year monthly means of Weare et al (1980).

This study is confined to one space-borne instrument. A number of satellite sensors can make similar observations. Wind speeds can be obtained from displacement of cloud images, from SMMR on NIMBUS-7, from SSM/I (Special Sensor Microwave/Imager) on DMSP and from altimeter on Geosat; the last two satellites are scheduled to be launched in 1984. Sea surface temperature can be measured by AVHRR (Advanced Very High Resolution Radiometer), HIRS (High Resolution Infrared Sensor) 2, and VAS (VISSR Atmospheric Sounder) on NOAA polar orbiting satellites and also from SMMR on NIMBUS-7. The integrated water vapor can be determined by SMMR, SSM/I, TOVS (Tiros Operational Vertical Sounder) and VAS. A combination of these observations that will optimize the estimation of latent heat flux should be found. But, first the performance of these sensors have to be rigorously compared and tested. The Sea Surface Temperature Workshops sponsored by the National Aeronautic and Space Administration and the Workshop on Interim Surface Wind Data Sets sponsored by the World Climate Research Program will provide the opportunities to compare and to improve various space-borne sensors.

This study is confined to a special region, but similar studies in the Tropical Pacific are being planned. The applicability of bulk parameterization in equatorial regions will also be evaluated. This study is paving the way to the application of space-borne sensors in studies of low-frequency variations of air-sea heat exchanges and global climate.

ACKNOWLEDGEMENTS

The author is indebted to Professor Pearn P. Niiler of the Scripps Institution of Oceanography for valuable suggestions and timely encouragements during this study. Discussion with colleague Robert Haskins prior to the study was also helpful. This study is performed at the Jet Propulsion Laboratory, California Institute of Technology under contract with the National Aeronautic and Space Administration.

REFERENCES

Anderson, R.J. and S.D. Smith, 1981, Evaporation coefficient for the sea surface from eddy flux measurements. J. Geophys. Res., 86, 449-459.

Bunker, A.F., 1976, Computations of surface energy flux and annual air-sea interaction cycles of the North Atlantic Ocean. Mon. Wea. Rev., 104, 1122-1140.

Chou, S.H. and D. Atlas, 1982, Satellite estimates of ocean-air heat fluxes during cold air outbreaks. Mon. Wea. Rev., 110, 1434-1450.

Deardorff, J.W., 1968, Dependence of air-sea transfer coefficients on bulk stability. J. Geophys. Res.,73, 2549-2557.

Esbensen, S.K. and R.W. Reynolds, 1981, Estimating monthly averaged air-sea transfer of heat and momentum using the bulk aerodynamic method. J. Phys. Oceanogr., 11, 457-465.

Fleagle, R.G. and J.A. Businger, 1963, An Introduction to Atmospheric Physics, Academic Press, New York, p. 62.

Katsaros, K.B., P.K. Taylor, J.C. Alishouse, and R.G. Lipes, 1981, Quality of Seasat SMMR (Scanning Multichannel Microwave Radiometer) atmospheric water vapor determinations, Oceanography from Space, J.F.R. Gower ed., Plenum Press, 691-706.

Large, W.G. and S. Pond, 1982, Sensible and latent heat flux measurements over the ocean. J. Phys. Oceanogr., 12, 464-482.

Lipes, R.G., 1982, Description of Seasat radiometer status and results. J. Geophys. Res., 87, 3385-3395.

Liu, W.T., 1981, Evaluation of parameterization models for determining air-sea exchanges in heat and momentum from satellite data. Application of Existing Satellite Data to the Study of the Ocean Surface Energetics, C. Gautier ed., University of Wisconsin Press, Madison, 35-41.

Liu, W.T. and K.B. Katsaros, 1983, Analysis of sea surface and air temperature from aircraft in the Joint Air-Sea Interaction (JASIN) experiment. J. Geophys. Res. (submitted).

Liu, W.T., K.B. Katsaros and J.A. Businger, 1979, Bulk parameterization of air-sea exchanges of heat and water vapor including the molecular constraints at the interface. J. Atmos. Sci., 36, 1722-1735.

Liu, W.T. and W.G. Large, 1981, Determination of surface stress by Seasat-SASS: A case study with JASIN data. J. Phys. Oceanogr., 11, 1603-1611.

Lorenz, E.N., 1977, An experiment in nonlinear statistical weather forecasting, Mon. Wea. Rev., 105, 590-602.

Lowe, P.R., 1977, An approximation polynomial for the computation of saturation vapor pressure. J. Appl. Meteor., 16, 100-103.

McPhaden, M.J., 1982, Variability in the central equatorial Indian Ocean, Part II: Oceanic heat and turbulent energy balances. J. Mar. Res., 40, 403-419.

Niiler, P.P., 1982, Tropic Heat - A Study of the Tropical Pacific Upper Ocean Heat, Mass, and Momentum Budgets, 30 pp.

Paulson, C.A., E. Leavitt and R.G. Fleagle, 1972, Air-sea transfer of momentum heat and water determined from profile measurements during BOMEX, J. Phys. Oceanogr., 2, 487-497.

Pond, S., G.T. Phelps, J.E. Paquin, G. McBean and R.W. Stewart, 1971, Measurements of the turbulent fluxes of momentum, moisture and sensible heat over the ocean, J. Atmos. Sci., 28, 901-917.

Smith, W.L., 1966, Note on the relationship between total precipitable water and surface dew point. J. Appl. Meteor., 5, 726-727.

Weare, B.C., P.T. Strub and M.D. Samuel, 1980, Marine Climate Atlas of the Tropical Pacific Ocean, U. of California, Davis, 253 pp.

WCRP, 1983, Interannual Variability of the Tropical Oceans and the Global Atmosphere (TOGA), The World Climate Research Programme, 55 pp.

THE DETERMINATION OF SURFACE FLUXES OF HEAT AND WATER BY SATELLITE MICROWAVE RADIOMETRY AND IN SITU MEASUREMENTS

Peter K. Taylor

Institute of Oceanographic Sciences, Wormley, Surrey, U.K.

For climate related large-scale oceanographic experiments the monthly mean heat flux over 500 km square ocean areas is assumed to be required to a target accuracy of $\pm 10 W/m^2$. Only in the North Atlantic and North Pacific Oceans do ships provide the 100 or more observations per month required. Satellite microwave radiometers could provide useful measurements of wind speed, sea surface temperature, midlatitude precipitation, and perhaps near surface humidity. However these values alone do not define the fluxes and the limited swath width means that, from the planned satellites, sampling will only at best be marginally adequate. A combination of satellite and _in situ_ data is therefore necessary; full exploitation of the satellite data will require _in situ_ measurements of improved absolute accuracy.

1. INTRODUCTION

During the operational period of WOCE and TOGA, passive microwave radiometers will be flown on a number of satellites. This paper will consider the potential contribution of these instruments to the determination of the vertical, turbulent fluxes of sensible heat, latent heat and water through the sea surface. Emphasis will be placed on what is likely to be achieved given the characteristics of the currently planned instruments and the orbit parameters of the satellites on which they will be launched.

There is an urgent need for sensitivity studies to determine, as a function of geographical position and season, the accuracy required for WCRP experiments. Pending further information it will be assumed for this paper that the monthly mean heat flux

C. Gautier and M. Fieux (eds.), Large-Scale Oceanographic Experiments and Satellites, 223–246.

over a 500km square region (or equivalent area) is required to $\pm 10W/m^2$. This is compatible with plans for the large-scale oceanographic experiments within the World Climate Research Programme, WCRP (CCCO 1982). For example, for the CAGE experiment, Dobson et al. (1982) defined a useful accuracy as 20% of the polewards heat flux which, for the North Atlantic and 24^oN, amounts to $\pm 10W/m^2$ for the annual mean flux. This is thought to be of similar order or greater than the interannual variability (Niiler & Richardson 1973, CCCO 1982). For the Tropical Ocean Global Atmosphere experiments, TOGA, ± 5 to $10W/m^2$ accuracy for monthly mean fluxes in boxes of 10° longitude by 2.5^o latitude is sought (CCCO 1982). Accuracy requirements of this order imply not only very small systematic errors in measurement, but also sampling adequate to reduce random errors, due to measurement error and smaller scale variability, to the required level.

Considering satellite observations for climate research likely to be available by the year 2000, COSPAR (1980) concluded that "we see no promising prospects for determining the flux quantities from ... oceans to the atmosphere except by very indirect means". Aircraft remote sensing of total heat flux is possible by using a two wavelength infrared radiometer to measure the temperature gradient in the top 0.1mm of the sea surface (McAlister & McLeish 1970, McAlister, McLeish & Corduan, 1971). However unknown atmospheric modulation of the received radiation makes application of this technique to satellite remote sensing difficult if not impossible. With the exception of budget techniques, and the method proposed by Atlas et al. (1981), which is applicable to the important but limited case of cold air flowing from land over coastal ocean, indirect flux determination relies on the bulk aerodynamic formulae (see, for example, Friehe & Schmitt 1976, Large & Pond 1982). These express the fluxes in terms of mean values of sea surface temperature and near surface air temperature, humidity and wind speed, all or some of which can be measured by ships, buoys or satellites.

The bulk aerodynamic equations have been used in previous heat flux studies by Budyko (1956), Bunker (1976), Bunker & Worthington (1976), Hastenrath & Lamb (1978), Weare, Strub & Samuel (1981) and others, who used data from the ships of the Voluntary Observing Fleet, VOF (WMO 1977). It will be necessary to consider the accuracy of these VOF observations in some detail for three reasons. (1) The VOF data represent the current state of knowledge against which the usefulness of the satellite estimates can be compared. (2) The satellites cannot measure all the required variables, a mixture of VOF and satellite data will be necessary. (3) Many of the microwave radiometer algorithms are semi-empirical and require <u>in situ</u> data for calibration and verification.

In the following sections the accuracy of the bulk formulae will be considered followed by the required sampling strategy. Then, for ships and for satellites, the adequacy of the sampling and the accuracy of the measurements will be discussed. It will be suggested that, to exploit fully the potential of satellites, continued and improved in situ measurements will be required.

2. ACCURACY OF THE BULK FORMULAE

Within the bulk aerodynamic formulae a number of physical processes are parametrized by the transfer coefficients, C_H and C_E for sensible and latent heat respectively. To the extent that the appropriate value of C_H and C_E can be determined for any particular situation, the bulk formulae will be accurate. C_H and C_E are functions of stability, measurement height and possibly wind speed and sea state; for recent discussions see Friehe & Schmitt (1976), Liu, Katsaros & Businger (1979), Anderson & Smith (1981), Large & Pond (1982). Recent experimental determinations, using the eddy correlation or dissipation methods (see Dobson, Hasse & Davies, 1980), give neutral values which are consistent to 10 to 15% (Large & Pond 1982) at least upto 15m/s wind. At higher wind speeds the production of spray might change the transfer coefficients but there is a lack of experimental data. Thus there is a need for more eddy correlation and dissipation estimates under a wide range of conditions, in order that the transfer coefficients be adequately known. One such experiment is the Humidity Exchange Over the Sea (HEXOS) programme (Smith, Katsaros & Oost 1983).

Unfortunately Bunker, Charnock & Goldsmith (1982) have demonstrated a further problem. The neutral values of C_E from experimental methods lie in the range 1.05 to 1.20×10^{-3}. However, in examining the Mediterranean and Red Sea heat budgets, they concluded that the appropriate value of C_E should be 1.8 to 1.9×10^{-3}. They ascribed the need for a high C_E value possibly to the ships' avoidance of high winds and systematic errors of observation. The discrepancy is not new, Robinson (1966) noted a similar difference between micrometeorological and climatological transfer coefficients. He believed that the cause was partly the use of mean values in the equations, an effect which Esbensen & Reynolds (1981) have suggested is not more than 10%, and inhomogeneity of the observational data base. Bunker et al. (1982) have also pointed to inhomogeneity in the data to explain differences in latent heat flux estimates by Goldsmith & Bunker (1979) and Hastenrath & Lamb (1979).

Unless the cause of these variations in the appropriate transfer coefficients can be understood, the correct values to use for evaluating fluxes from a mix of satellite and VOF data will not be known. Furthermore, if the satellite values have been calibrated with respect to VOF ships, higher transfer coefficients may be needed than for calibration against research ships. Clearly more research into the accuracy and representivety of the VOF reports is needed.

3. FREQUENCY OF SAMPLING

3.1 Required sampling density

The sampling density will be defined by the number, n, of observations during one month in a 5° square of ocean (equivalent to a 500 km square area within the accuracy of the following estimates). The adequate value of n for surface flux determination will depend on the spatial and temporal variability in the atmosphere and ocean and will therefore be a function of geographical location and season. Fissel, Pond & Miyake (1977) suggest that n>15 would be adequate to define sensible heat and momentum fluxes at O.W.S. Papa (50^oN) provided the estimates were evenly spaced and, somewhat surprisingly, a smaller value of n would suffice for latent heat. However this estimate excludes the effects of spatial variability, and measurement errors. Weare & Strub (1981) considered the effect of sampling biases, given diurnal and spatial variability, on calculated monthly mean fluxes but their analysis was limited to two coastal regions in the tropical Pacific. They suggest n>11 for such regions but quote N.E. Clark (of Clark 1967) as suggesting that n=50 to 75 would be sufficient at OWS Papa during Winter. Dobson et al. (1982) state that T.P. Barnett (1981, unpublished) found that at least 100 VOF observations during a month were necessary to give sampling of storm events comparable to a nearby weathership, that is n>100. Gross errors resulted when less than 40 observations were available. Harrison (1983) reported that 20 to 30 samples per month were necessary in order to attain a 0.9 correlation between reconstructed and true time series in the central equatorial Pacific.

From this sparse evidence one might guess that for much of the world ocean n=20 would be barely adequate while n>100 would give reasonable heat flux estimates in the absence of significant measurement error. A limited check of these criteria can be made using data from the Joint Air Sea Interaction Experiment, JASIN (Pollard, Guymer & Taylor 1983). For this midlatitude (59^oN 12^oW) summertime experiment the fluxes were small, $21W/m^2$ was the mean latent heat flux based on hourly sampling. Figure 1 shows the increased uncertainty in defining this flux if meteorological measurements were only available at less frequent intervals. For example sampling at about 30 hours' period is necessary to define the latent heat flux to $\pm 4W/m^2$, that is 20% of the mean (n>24).

For sensible heat flux 5 hour sampling would be necessary (n~140). However these conclusions only apply to this region during this period. More general conclusions could be reached by using weather ship data.

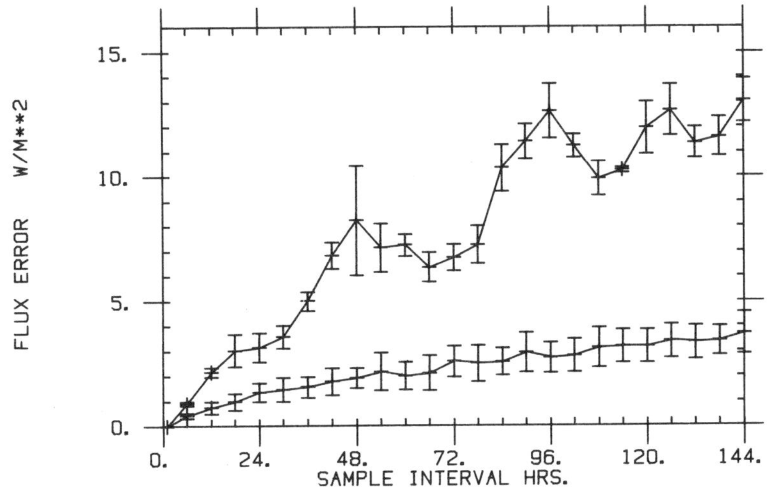

Figure 1. Uncertainty in estimating the surface fluxes for phase 2 of the JASIN experiment due to sampling at increased time intervals compared to hourly sampling. Upper curve: latent heat flux; lower curve: sensible heat flux.

3.2 Sampling by ships

Values of n during FGGE have been tabulated by WMO/ICSU (1981a,b,c). Assuming that the density at the start of FGGE will be typical for WCRP, it is possible to derive the values shown in table 1 for representative areas of the oceans. Only in the North Atlantic and possibly North Pacific Oceans was the sampling adequate; in the tropical Atlantic Ocean it was barely adequate. During FGGE the sampling in data sparse areas was increased; however some of this apparent improvement was due to the FGGE drifting buoy programme which only provided pressure and SST values. If a similar improvement were attained for WCRP, but with full observations from the buoys, then the sampling would be barely adequate in the tropical Atlantic and Indian Oceans but remain inadequate in the South Atlantic, Tropical and South Pacific, and Southern Oceans.

Table 1. Number of observations per month in 5° squares of ocean during the period 27 Nov. '78 - 5 Jan. '79 (WMO/ICSU 1981). Mean, maximum and minimum values are given for representative areas of each ocean. Also shown is the percentage, P, of 5° squares for which n>100, n>20, and (*last column) n>20 given the sampling improvement which occurred during FGGE (a factor of 2.5 in the Southern Ocean, 1.2 in tropical areas). Significant values are underlined.

Ocean	Area	n			P%		
		Mean	Max	Min	n>100	>20	>20*
Atlantic	North	207	1011	18	<u>79</u>	<u>98</u>	-
	Tropical	54	216	0	<u>19</u>	<u>76</u>	<u>80</u>
	South	63	345	0	30	<u>45</u>	<u>63</u>
Pacific	North	153	768	24	<u>68</u>	<u>100</u>	-
	Tropical	57	759	0	<u>23</u>	44	50
	South	48	312	0	21	41	53
Indian	Tropical	42	1680	3	15	61	<u>75</u>
	South	33	360	0	6	48	<u>76</u>
Southern	-	24	390	0	10	23	31

It must also be stressed that these calculations are based on the FGGE level 2b data set (WMO/ICSU 1978) which includes data obtained from ships logs. Weare & Strub (1981) found a factor of 8 decrease in reports at 4am local time if radio reports reaching the Global Telecommunications System alone were used. The level 2b data set assembly required a special effort during FGGE and it may be necessary to implement a similar effort during WCRP to achieve similar sampling levels.

3.3 Sampling by satellites

Table 2 shows the orbit characteristics for satellites which will carry microwave radiometers during the WRCP period. Table 3 shows details of the radiometers. There are three important points. First, the potential number of available instruments is very small, and the only firmly planned satellites carrying multichannel radiometers capable of measuring a wide range of

Table 2. Satellites planned or proposed for the period of the WCRP which will carry microwave radiometers.
Key to acronyms: DMSP - Defence Meteorological Satellite Programme (U.S.A.); MOS-1, MOS-2 - Marine Observational Satellites (Japan); ERS-1 - Earth Resources Satellite (Europe); NROSS - Navy Remote Ocean Sensing System (U.S.A.); NOAA-H, I - U.S.A. operational meteorological satellites.
Status (1) Planned, (2) Proposed (3) Tentative.

Satellite	Op. Period	Ht. (km)	Inclination (deg)	Period (mins)	Repeat (days)	Status
DMSP (5D-2)	1985 to 1991	833	98.7	101	~4	(1)
MOS-1	1988-1991	909	99.1	103	17	(1)
ERS-1	1988-1991	777	98.5	100	3	(1)
NROSS	1988?	830	99	102	?	(2)
POSEIDON	1988?	832 or 1100	98	101?	26?	(2)
NOAA-H NOAA-I	1987	830		102	-	(1)
MOS-2	1990?	-	-	-	-	(3)

parameters are DMSP and NROSS. Both of these are operational, military satellites, and it is not currently planned to calculate atmospheric water vapour, a climatologically useful variable, from the DMSP data. Secondly, most microwave radiometers scan a limited swath, considerably reducing the frequency of sampling compared to the limb to limb scan of, for example, a visible and infrared radiometer. Thirdly, in order to achieve reasonable resolution given the available antenna size, and to allow sea ice determination, the microwave radiometers are mounted on polar orbiting satellites. This imposes further sampling restrictions.

Salby (1982a,b) has shown that, at best, sampling by a nadir

Table 3. Swath, frequencies and resolution of microwave radiometers proposed for satellites in table 2. Acronyms: SSM/I - Special Sensor Microwave/Imager MSR - Microwave Scanning Radiometer; (MR) - unspecified microwave radiometer; MSU - Microwave Sounding Unit; AMSU - Advanced Microwave Sounding Unit.

Microwave Radiometer	Sat.	Swath (km)	Frequ. GHz	Res. (km)	Recovered Parameters
SSM/I	DMSP	1394	19.35V+H	55	Surface Wind
			22.2V	49	Rain
			37 V+H	33	Cloud Water
			85.7V+H	15	Liquid Water
MSR	MOS-1	320	23.8	32	Atmos. water
	MOS-2?		31.4	23	Sea ice
ATSR/M	ERS-1	Nadir	23.8	22	Atmos. water
			36.5		
(MR)	NROSS	(1) Similar to SSM/I?			As SSM/I plus sea temperature
		(2) Low frequency MR			
(MR)	Poseidon	Nadir?	?	?	Atmos.water?
MSU	NOAA-H	2250	50.3	110	Vertical, temperature profiles
	NOAA-I		53.7		
			55.0		
			58.0		
AMSU	Later NOAA (1990's)	2320	20 channel 20-183 GHz	~50	Vertical sounding instrument (proposed).

sounding polar orbiting satellite is equivalent to fixed point sampling once per day at widely spaced points. Considering the SSM/I with a 1400 km swath width, in one day the area scanned is approximately equal to 1.5 times the surface of the earth (fig. 2). With two satellites in suitable orbit the whole earth can be sampled twice per day with restricted regions sampled 4 or more

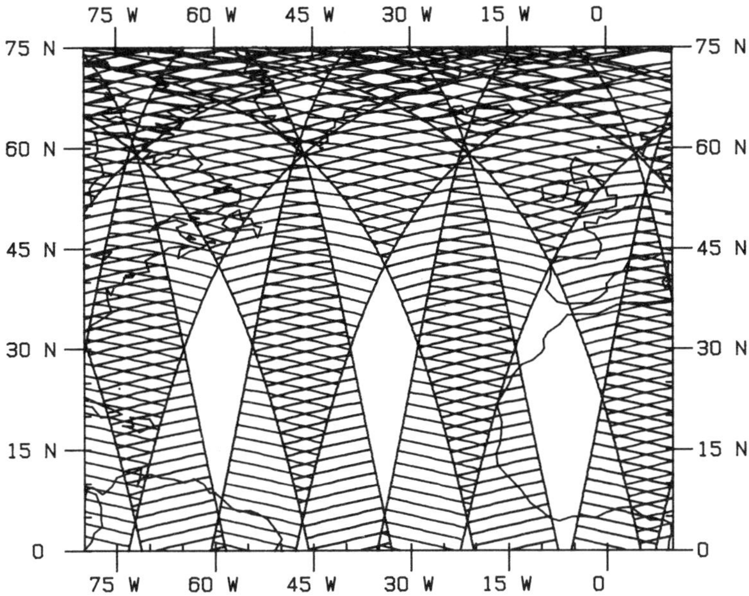

Figure 2. Area of the North Atlantic Ocean sampled by the SSM/I on a DMSP satellite during one day.

times. This is equivalent to m=60 to 120 for most regions, marginally adequate for well defined monthly mean values. The sampling provided by MOS-1, ERS-1 and Poseidon will be much worse. Only if DMSP and NROSS are simultaneously in orbit, or if the proposed AMSU (RAL 1982a) comes into being in the 1990's, will microwave radiometers begin to provide the coverage sought. Even then, processes with a marked diurnal signal, such as tropical precipitation (section 4.4), will be poorly resolved.

4. ACCURACY OF THE MEASUREMENTS

4.1 Sea Surface Temperature (SST)

Target accuracy. The most stringent requirement is for latent heat flux in the Trade wind or Tropical regions (SST ~25°C) where an SST error of 0.2 to 0.3°C will produce a $10W/m^2$ flux error for a 10m/s wind. For midlatitude areas (SST ~10°C), and for sensible heat flux, SST is required to about ±0.7°C for similar wind speed and flux error.

Ship measurements. While it is difficult to determine the absolute accuracy of ship measurements, some information is provided by the study of James & Fox (1972) which compared the two most widely used measurement techniques, bucket and engine intake temperatures. The latter were biased high by 0.3°C compared to bucket values. 87% of bucket/intake value pairs agreed to within 2.9°C, 68% agreed to within 0.9°C.

Satellite measurements. Satellite SST measurements are considered in detail by Minnett et al. (1983) and Bernstein (1983). Microwave instruments such as the Seasat SMMR have achieved better than ±1°C under favourable conditions (Lipes, 1982, Taylor, 1983a). However careful screening is necessary to remove the effects of sunglint, radio-frequency interference, and heavy rain and retrievals cannot be made within 600km of land. Future infrared instruments, such as the ATSR (RAL, 1982b) are expected to achieve accuracies within a few tenths °C, but sampling problems related to cloud cover will remain. The only future microwave radiometer designed to measure SST is that planned for NROSS. If either microwave or infrared methods are to attain the high target accuracy, calibration against high quality in situ measurements will be necessary (e.g. Strong [1982]) with appropriate correction for the effect of the cool ocean surface skin (Robinson, Wells & Charnock, 1983).

4.2 Wind speed

Target accuracy. The fluxes are proportional to wind speed so a target accuracy of ±10% or less should be sought.

Ship measurements. Probably over 80% of all VOF ship wind reports are estimates, by the ships officers, of the Beaufort wind force based on the sea state; with factors such as the handling of the ship probably contributing, particularly at night. The British Meteorological Office consider this method preferable to the use of anemometers (see, for example, Shearmann, 1983) because the sea state reflects the near surface wind integrated over perhaps a one hour period. There are also difficulties in installing anemometers with unobstructed exposure on merchant vessels and the increased instrument cost may be hard to justify for weather forecasting purposes (WMO, 1970). For example, Verploegh (1967) found a mean uncertainty of only ±1.3m/s for estimated winds compared to ±1.0m/s for anemometers. However Graham (1982) found that the standard deviation of VOF estimated winds compared to measured values increased from about ±2m/s at force 1 to ±5m/s at force 9. This compares poorly with the differences between well exposed anemometers on research ships during JASIN, which Macklin & Guymer (1980) found to be ±0.75m/s or less.

A further uncertainty in estimated winds arises from the scale

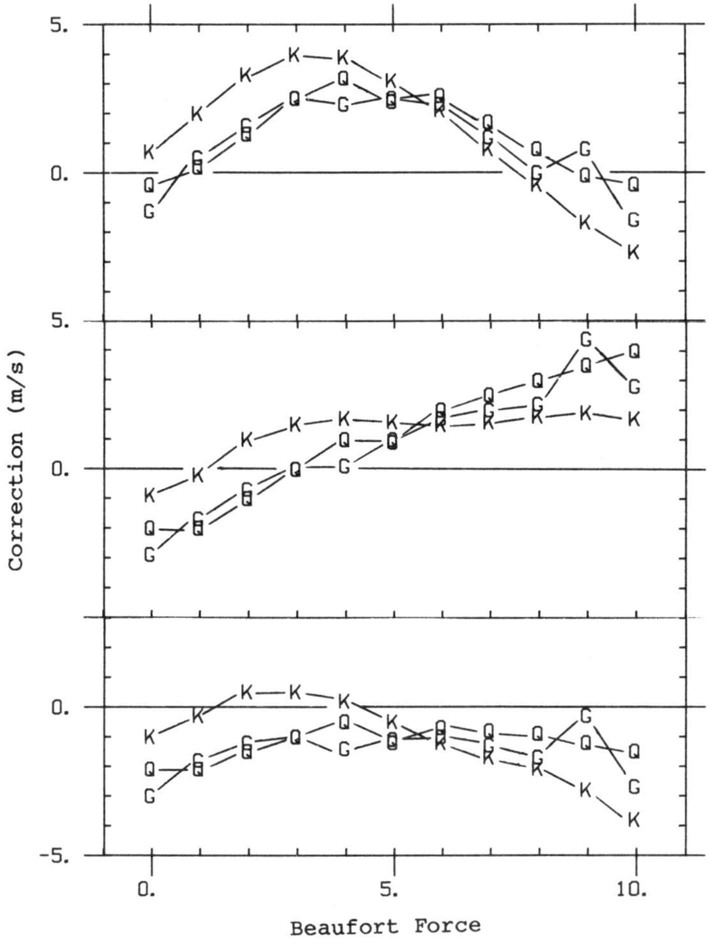

Figure 3. Correction required to the standard Beaufort wind scales to give agreement between estimated winds and those measured by nearby weatherships. Data is from Q=Quayle (1980), K=Kaufeld (1981) and G=Graham (1982). Top: correction for Code 1100. Middle: correction for CMM IV. Bottom: for CMM IV if weathership winds were first amended using u'=0.89u+0.3 (m/s).

used to convert Beaufort force to m/s. The WMO recognized scale, Code 1100 (WMO 1970), was adopted in 1946 and is still used by ships' officers in coding their wind reports. However a revised scale, CMM IV, based on substantially more extensive data (Verploegh, 1956, 1967, Frost 1966) has been adopted by WMO for scientific use. Recent statistical studies of the difference between VOF and weather ship winds have been reported by Quayle (1980), Kaufeld (1981) and Graham (1982). Figure 3 (top) shows the correction needed to Code 1100 to bring the estimated winds into agreement with the measured winds. Code 1100 appears to underestimate the wind speed in the most commonly observed region, force 3-5. In contrast CMM IV (figure 3 middle) appears to overestimate at low wind speed and underestimate at high values. However this assumes that the measured values are correct, whereas the possibility of error can be illustrated by an example from the JASIN experiment. Comparison of one ship, which used an anemometer and dial system of the type used on British weather ships, to other ships which used light weight research type anemometers and electronic averaging, showed that the Weather ship type needed both a starting speed correction and a 10% overspeeding reduction. Figure 3 (bottom) shows the effect of arbitrarily applying these corrections to the data. Using CMM IV the estimated winds are now in closer agreement with the measurements, although a bias remains. Unless it can be determined whether it is the Weather ship observations or the CMM IV scale that is in error, there will always be the possibility of a 10% systematic bias in VOF winds. A similar bias would occur in satellite winds calibrated with respect to the biased _in situ_ data. The decreasing number of weather ships means that research into the accuracy of past weather ship observations is an urgent need. For example, a meteorological buoy could be moored in the vicinity of a weather ship and the observations compared.

Satellite measurements. In common with the scatterometer (Jones et al. 1982, Guymer 1983) and altimeter (Brown 1979, Fedor & Brown 1982), the Seasat SMMR achieved wind speed accuracies of ±2m/s or better (Lipes 1982, Taylor 1983). Away from areas of sunglint and at least 600 to 700 km from land the SMMR was in good agreement with the scatterometer and the SMMR winds showed the least scatter compared to _in situ_ measurements (Wentz, Cardone & Fedor 1981). However the evaluation was for a limited range of wind speeds and atmospheric states, and there were possible calibration changes with time (Allan & Guymer 1983) suggesting further evaluation is necessary. Thus although microwave radiometers may be capable of providing useful wind measurements, calibration and verification against good _in situ_ data will be necessary. This will be particularly true for SSM/I which does not have the lower frequency channels used by SMMR for wind determination.

4.3 Air temperature and humidity

Target accuracy. In general, for $10W/m^2$ heat flux error for a 10m/s wind and at $25^{\circ}C$ air temperature, both dry and wet bulb temperatures must be known to better than $\pm0.2^{\circ}C$. The equivalent precision for specific humidity being ±0.3g/kg. However if changes of C_E and C_H with stability can be ignored, for example for neutral or unstable conditions, then errors in determining the dry bulb temperature produce nearly equal and opposite changes in the sensible and latent heat flux. Thus, in this case, only the wet bulb temperature is necessary to define the total heat flux.

Ship measurements. The deficiencies in ship temperature and humidity observations are well known and the possibility of a systematic bias has been mentioned above (section 3). Salt contamination of the wet bulb, or heat island effects due to the ship, will increase the wet bulb temperature and apparent specific humidity and result in too low a value for the latent heat flux from the sea to the air. Heat island errors will be particularly large if the true wind velocity and ships velocity are similar with errors of a few degrees possible (Robson 1973). Knowledge of the relative wind would allow some quality control but only the true wind is transmitted. The only ships velocity information recorded refers to an average over the previous three hours, provided to allow interpretation of the barometric tendency.

Satellite measurements. Current or planned satellite instruments for measuring the atmospheric profiles of temperature and humidity do not approach the required accuracy for estimates of near surface air temperature and humidity (see, for example COSPAR 1980). Indirect methods must therefore be considered.

The success of microwave radiometers such as SMMR in defining the total atmospheric water content, W (e.g. Taylor, 1983a), has prompted attempts to so determine the latent heat flux, F_E. Budget techniques have been proposed (Pierson 1978, Ruttenberg 1981), but these require knowledge of the vertically integrated horizontal humidity tranport and this is not the same as the product of the mean humidity and mean wind.

Another possibility is to make use of the correlation betwen W and near surface specific humidity q_a (see, for example, Reitan 1963, Tomasi 1981). This has been discussed at this conference by Liu (1983), who has obtained ±35 W/m^2 accuracy for mean monthly latent heat fluxes. However, Taylor (1982, 1983b) has suggested that the variability of atmospheric structure may be such as to prevent a substantial improvement in this accuracy. This is because, for ocean areas remote from land, the near surface air temperature and humidity reflects the values of SST along the air trajectory as much as the humidity of the upper atmosphere. Indeed, the correlation between q_a, calculated by assuming a constant relative humidity (say 80%) for the near surface air, and the observed q_a values, may be as good as the correlation between q_a and W (Taylor 1983b). Using a set of radiosonde profiles for different places and seasons, the predicted uncertainty in flux determination is equivalent to about ±80W/m^2 for a 10m/s wind, which, given a typical synoptic period of 2-3 days, is unlikely to be reduced below ±25W/m^2 for a monthly mean value. Nevertheless, this technique may still be the best available, either by satellite or ship, for many ocean areas and further evaluation of the method is therefore worthwhile. For example a spatial filter applied to the W values might reduce the scatter due to frontal moisture, or the method might be used in combination with satellite humidity profiles or ship measurements.

4.4 Precipitation

Target accuracy. The desired accuracy quoted by CCCO (1982) represents the water flux equivalent to ±10W/m^2 latent heat flux, that is ±0.35 mm/day, or of order 5% of the annual precipitation. However one use of the precipitation value is to determine the net moisture flux, which affects the salinity of the upper ocean, and thus may precondition water masses for overturning. For these purposes it is the density change which is important and ±10W/m^2 heat flux is equivalent, in terms of density change, to about ±1.2mm/day for typical upper ocean temperatures and salinities (i.e. ±20% of the annual precipitation rate). Unfortunately the difficulty of estimating precipitation rate over the ocean is well known (Atlas & Thiele 1981, Austin & Geotis 1980, WMO 1976, 1969, 1962) and even ±20% accuracy may be a difficult target.

Nature of global rainfall. The effectiveness of satellite and in situ methods for determining precipitation varies due to the different nature of precipitation in the tropics compared to the westerly wind zones. The tropical region will be defined as the area equatorward of the subtropical high pressure areas. The tropical rainfall is generally convective in nature and therefore spasmodic. The cumulonimbus clouds may be organized into mesoscale systems with maximum dimensions of a few 100km, although within these systems the rain cells are of order a few km in diameter (Ruprecht & Gray 1976). For example, during GATE two events produced one third of the total precipitation, and 50% of rainstorms delivered 75% of the precipitation due to such systems (Riehl 1979). Precipitation rates are correspondingly high. In GATE the instantaneous rainfall rate within a 4km square area was greater than 3mm/hr on 50% of rainfall occasions and greater than 5mm/hr for 35% of such occasions (Hudlow & Patterson 1979). Such precipitation often shows a marked diurnal variation (Gray & Jacobson, 1977). Wind speeds in the tropics are generally less than 10m/s and waves low, except in larger organized systems such as tropical cyclones.

Although annual precipitation is a maximum in the tropics at about 2 to 3m, it is of a similar order of magnitude, generally 1 to 2m in the westerly wind region (all these figures have at least ±50% uncertainty). In the westerlies the slight or moderate continuous rain categories contribute a significant fraction of the rainfall (Britton 1976). Such rainfall is likely to be geographically extensive (order 1000 km although it is often banded with a length scale of 100-200km, see reviews of Browning & Mason (1980) or Hobbs (1978, 1981)) and persistent, being associated with frontal systems which evolve during a period of days. Mean rainfall rates are generally less than 3mm/hr (Britton 1976). Rainfall is associated with increased wind and wave conditions.

Ship measurements. Ships navigational radar is designed to minimize the signal from rain and is not suitable for rainfall measurement (WMO 1976). The methods available for determination of rain from VOF ships are either by rain gauge or by inference from the present and past weather report. Austin & Geotis (1980) and WMO (1962) review the difficulties of siting raingauges in ships. Early experiments (WMO 1962) had suggested that raingauges should be positioned as high as possible but trials were generally discouraging (e.g. Robson 1973) with possible errors of a factor of 2 or more. These, however, were Westerly region trials and ship motion and airflow disturbance were major factors. In the tropics during GATE these factors were absent and gauges in a single ship agreed to within 12% of the total (Austin & Geotis, 1980). Unfortunatly the sporadic nature of tropical precipitation means that accurate rainfall measurements, even from an array of points as in GATE, have limited value for monthly rainfall determination (Riehl 1979). For WCRP purposes the most important use of shipboard gauges will be to serve as calibration points for satellite measurements; a similar conclusion was reached by Austin & Geotis (1980). For this purpose improved raingauge measurements in westerly wind regions, making use of ship flow distortion studies, are highly desirable.

The use of VOF ships's weather reports to estimate rainfall (Tucker 1961) was considered in detail by Britton (1976) who concluded that rainfall amounts for mid-latitude areas were underestimated. Extensions of the method by various means to other areas have been used by Dorman & Bourke (1981) and Reed & Elliott (1979) to obtain values which differ by a factor of 2 or 3 even in the relatively well sampled Tropical and North Atlantic Ocean. However the extensive data base available from the VOF ships suggests that the method should continue to be developed.

Satellite measurement. COSPAR (1980) and Atlas & Thiele (1981) have reviewed satellite methods for precipitation measurement. COSPAR (1980) concluded that monthly rainfall estimates are attainable over the oceans with accuracies of the order of ±20% for areas of about 10^5km^2 from one or more visible/IR methods. However these methods, often relying on geostationary satellites, have been developed mainly for tropical areas. Application to the stratiform rain characteristic of westerly wind regions may not produce the accuracy quoted.

Microwave radiometers provide an estimation of precipitation rate. Advantage has been taken of the detailed precipitation measurement available during GATE to calibrate rainfall measurements from the Nimbus 5 Electrically Scanning Microwave Radiometer, ESMR, (Austin & Geotis 1980). Unfortunately the large footprint of a microwave instrument, the poor tropical sampling provided by a single polar orbiting satellite, and the nature of tropical rainfall limit the accuracy of the estimates (Lovejoy & Austin 1980). ESMR was found to underestimate the rainfall on average by about 40% compared to radar and gauge measurements with standard deviation of order 80% (Austin & Geotis 1980). Corrections to the ESMR estimates have been proposed (Ruprecht, Medrow & Raschke 1983) but an underestimate remains and the standard deviation is increased. Empirical calibrations may be necessary to obtain monthly mean values.

The nature of westerly wind region rainfall is more suitable for microwave measurement. The Seasat Scanning Multichannel Microwave Radiometer (SMMR) provided maps of frontal precipitation (Katsaros, Taylor, Alishouse & Lipes 1980, Taylor, Guymer, Katsaros & Lipes 1983, McMurdie 1983, Taylor 1983a) but adequate surface measurements were generally not available. Evaluation of the accuracy of Nimbus 7 SMMR rainfall measurements is continuing (Katsaros & McMurdie 1983). Some critical assumptions needed in interpretation of passive microwave measurements could be relaxed if rainfall parameters were also estimated using a radar altimeter (Goldhirsh and Walsh 1981) on the same satellite. Since a microwave radiometer also supplies height correction values for the altimeter this combination of instruments may fly on a number of satellites and efforts should be made to ensure that the altimeter is capable of rainfall estimation in addition to its primary role.

6. DISCUSSION

Considering first precipitation, each method of ocean rainfall determination has advantages and disadvantages and, while different methods work best for different areas, in no case is a reliable absolute calibration available. It will be necessary to use a selection of methods perhaps combined using a General Circulation Model (GCM). For example, GCM estimates of atmospheric water flux divergence, together with surface latent heat flux measurements, could provide a consistency check on the rainfall estimate. Present GCM flux divergences, however, are not adequate for this purpose (Dobson et al. 1982). Microwave radiometers will be potentially most useful in defining extratropical rain rate. They will also provide atmospheric water vapour measurements which could be used as a check on the GCM predicted values. These radiometer rain rate estimates would benefit from improved rainfall measurement on ships in the westerly wind areas.

Indeed for each of the variables which can be measured by microwave radiometers (precipitation, sea surface temperature, wind speed, and possibly near surface humidity) full exploitation of the satellite data requires improved in situ measurements. Suggestions for improved V.O.F. instrumentation are not new (e.g. Vockeroth & Cook 1973), and ESSA (1969) has shown that so equipping a suitable subset of between 120 to 300 out of the 7500 V.O.F. ships would provide reasonable data coverage over the northern hemisphere at least. With careful anemometer siting, the use of electronic wind velocity averaging, and improved ship navigation, the resulting data set would avoid the uncertainty in the Beaufort scale values and provide calibration points for the satellite measurements. Improved temperature and humidity sensors, sited according to the results of flow distortion studies, would minimize the uncertainty in the transfer coefficients C_E and C_H. In this respect, measured winds should be reported in terms of relative wind and ships velocity. This would both avoid errors in calculating the true wind and also improve quality control of the winds, temperatures and humidities. The new observation code (Stubbs 1981), does not allow for this but provision is made for extra groups to be inserted.

It has been suggested by Large & Pond (1982) that the bulk aerodynamic formulae be expressed in terms of the friction velocity, u_*, thus reducing the stability dependence and accounting for any variation on the roughness length, z_o. Variations between bulk aerodynamic and dissipation momentum fluxes of up to 30% were found by Large & Pond (1981) for averaging periods of up to one day and attributed to effects of changing wind velocity. However for monthly averages the difference is likely to be less than 3% and correction of monthly bulk estimates is probably not necessary. Microwave radiometer (and scatterometer) estimates of wind speed may, however, be sensitive to z_o and this could lead to errors when calibration is performed against a limited number of spot in situ measurements. One solution is for at least some ships to measure stress using the inertial dissipation technique (e.g. Dobson, Hasse & Davis 1980). The required sensor, for example a Gill anemometer (Pond, Large, Miyake & Burling 1979) is considered suitable for operational use. Automatic stress calculation is electronicaly feasible but instrumentation would have to be developed for operational use.

7. CONCLUSIONS

(a) The estimation of monthly mean sensible and latent heat flux values on a 500km scale over the worlds ocean will continue to rely on the bulk aerodynamic formulae.

(b) The major source of *in situ* measurements for the bulk method is the ships' weather reports from the Voluntary Observing Fleet (VOF). If it were possible to reduce errors in these reports to a negligible level then at best the sampling appears to be adequate to define heat fluxes in the North Atlantic and North Pacific Oceans.

(c) Very high measurement accuracy is required if systematic errors in the flux estimates are to be reduced to an acceptable level. Further experiments are needed both to determine the transfer coefficients under a wider range of conditions and to identify and correct the systematic errors in measurements from the V.O.F. ships. The resulting flux estimates must be checked, for example by budget studies for semi-enclosed seas.

(d) Satellite instruments will not provide all the measurements required to define the fluxes. Microwave radiometers may provide useful measurements of surface wind and sea surface temperature for use with the bulk aerodynamic formulae. Satellite estimation of near surface air temperature and humidity used alone will not be of adequate accuracy for bulk determinations.

(e) A mixture of satellite and *in situ* data is therefore required. The necessary interpolation would best be done under the physical constraints imposed within an atmospheric General Circulation Model. The use of GCM's for this purpose will, however, require further development.

(f) Full exploitation of the satellite measurements will not be possible without improvement of the *in situ* measurements. An improved meteorological measurement package should be developed and mounted on a carefully chosen subset of V.O.F. ships. Data collection and relay by satellite would be highly desirable.

8. REFERENCES

Allan, T.D. & Guymer, T.H. (1983) Seasat measurements of wind and waves on selected passes over JASIN (accepted by Int. J. Rem. Sen).

Anderson, R.J. and S.D. Smith (1981) Evaporation coefficient for the sea surface from eddy flux measurements. J. Geophys. Res. 86, 449-456.

Atlas, D., Chou, S.-H. Hwang, P. and Stage, S.A. (1981) Estimating ocean-air heat exchange during cold air outbreaks from satellite observations. Presented at IAMAP 3rd Scientific Assembly, Hamburg, August, 1981.

Atlas, D. and O.W. Thiele (Eds.) (1981) Precipitation measurement from Space, Report, Goddard Laboratory for Atmospheric Sciences, NASA, Greenbelt (USA), 358 pp.

Austin, P.M. and S.G. Geotis (1980) Precipitation measurements over the ocean. In: Air-Sea Interaction: Instruments and methods (F. Dobson, L. Hasse, R. Davis Eds.) Plenum Press, New York & London.

Bernstein, R. (1983) Satellite sea surface temperature determination from microwave and infrared radiometry, (this volume).

Britton, G.P. (1976) Assessment of rainfall at sea - Feasibility study, in WMO (1976) pp 23-65.

Brown, G.S. (1979) Estimation of surface wind speeds using satellite-borne radar measurements at normal incidence, J. Geophys. Res. 84, 3974-3978.

Browning, K.A. & Mason, B.J. (1980) Air motion and precipitation growth in frontal systems, Pure appl. Geophys. 119, 577-593.

Budyko M.I. (1956) The heat balance of the Earth's surface, Leningrad: Hydrometeorological Publishing House, 254 pp (in Russian).

Bunker, A.F. (1976) Computations of surface energy flux and annual air-sea interaction cycles of the North Atlantic Ocean. Monthly Wea. Review, 104, 1122-1140.

Bunker, A.F. and L.V. Worthington (1976) Exchange energy charts of the North Atlantic Ocean. Bulletin of the American Meteorological Society, 57, 670-678.

Bunker, A.F., H. Charnock and R.A. Goldsmith (1982) A note on the heat balance of the Mediterranean and Red Seas. J. Mar. Res. (supplement), 73-84.

CCCO 1982 Large-scale oceanographic experiments in the World Climate Research Programme. Report of study conference, Tokyo, 10-21 May 1982.

Clark, N.E. (1967) Report on an investigation of large-scale heat transfer processes and fluctuations of sea-surface temperature in the North Pacific Ocean, Ph.D. dissertation, Mass. Inst. of Technology.

COSPAR (1980) Space based observations in the 1980s and 1990s for climate research: a planning strategy, Conf. Proc.

ECMWF, 24-28 March 1980, ICSU/COSPAR.

Dobson, F.E., F.P. Bretherton, D.M. Burridge, J. Crease, E.B. Kraus and T.H. Vonder Haar (1982) The CAGE Experiment: a feasibility study. WCP-22, WMO Geneva, 95 pp.

Dobson, F., L. Hase and R. Davis (Eds.)(1980) Air-Sea Interaction: instruments and methods. Plenum, New York, 801 pp.

Dorman, C.D. and R.H. Bourke (1981) Precipitation over the Atlantic Ocean, 30°S to 70°N. Mon. Wea. Rev. 109, 554-563.

ESSA (1969) Traffic aspects of merchant ships as environmental observing platforms, Weather Bureau Technical Memorandum 5PDD2, U.S. Dept. of Commerce (reprinted in WMO 1970b).

Esbensen, S.K. and R.W. Reynolds (1981) Estimating monthly averaged air-sea transfers of heat and momentum using the bulk aerodynamic method. J. Phys. Oceanogr. 11, 457-465.

Fedor, L.S. and Brown, G.S. (1982) Waveheight and wind speed measurements from the Seasat Radar Altimeter, J. Geophys. Res. 87, 3254-3260.

Fissel, D.B., Pond, S. and Miyake M. (1977) Computation of surface fluxes from climatological and synoptic data, Mon. Weath. Rev. 105, 26-36.

Friehe, C.A. and K.F. Schmitt (1976) Parameterizations of airsea interface fluxes of sensible and latent heat by bulk aerodynamic formulae. J. Phys. Oceanogr. 6, 801-809.

Friehe, C.A. and Pazan, S.E. (1978) Performance of an air-sea interaction buoy, J. Appl. Meteorol. 17, 1488-1497.

Frost, R. (1966) The relation between Beaufort force wind speed and wave height, Meteorological Office Scientific Paper No. 25, H.M.S.O. London 12 pp.

Goldhirsh, J. and E.J. Walsh (1981) Precipitation measurements from space using a modified Seasat type radar altimeter. Johns Hopkins University. Applied Physics Lab., No. S1R8IU-022, 77 pp.

Goldsmith, R.A. and A.F. Bunker (1979) Woods Hole Oceanographic Institution collection of climatology and air-sea interaction data. Woods Hole Oceanographic Institution, Technical Report No. WHOI-79-70.

Graham, Anne E. (1982) Winds estimated by the Voluntary Observing Fleet compared with instrumental measaurements at fixed positions, Met. Mag. 111, 312-327.

Gray, W.M. and Jacobson, R.W. (1977) Diurnal variation of deep cumulus convection, Mon. Weath. Rev. 105(9), 1171-1188.

Guymer, T.H. (1983) A review of the Seasat-A satellite scatterometer, Phil. Trans. Roy. Soc. Lond. A279 (in press).

Harrison, D.E. (1983) New directions in surface fluxes, I.U.G.G. Inter-disciplinary Symposia, Abstracts vol. 2, 804.

Hastenrath, S. and P. Lamb (1978) Heat budget atlas of the Tropical Atlantic and Eastern Pacific Oceans. University of Wisconsin Press, 104 pp.

Hastenrath, S. and P. Lamb (1979) Climatic atlas of the Indian

Ocean. University of Wisconsin Press, 104 pp.
Hobbs, P.V. (1978) Organization and structure of clouds and precipitation on the mesoscale and microscale in cyclonic storms. Rev. Geophys. Space Phys. 16, 741-755.
Hobbs, P.V. (1981) Mesoscale structures in midlatitude frontal systems, in: Nowcasting: Mesoscale observations and short range prediction, E.S.A. SP-165, 29-36.
Hudlow, M.D. and Patterson, V.L. (1979) GATE Radar Rainfall Atlas, NOAA Special Report.
James, R.W. and Fox, P.T. (1972) Comparative sea-surface temperature measurements, Reports on Marine Science Affairs No. 5. (WMO-336).
Jones, W.L., L.C. Schroeder, D.H. Boggs, E.M. Bracalente, R.A. Brown, G.J. Dome, W.J. Pierson, F.J. Wentz (1982) The SEASAT-A Satellite Scatterometer: The geophysical evaluation of remotely sensed vector winds over the ocean. J. Geophys. Res. 87, 3297-3317.
Katsaros, K.B. and McMurdie, L.A. (1983) Atmospheric water distribution in cyclones as seen with scanning multichannel microwave radiometers (SMMR). Int. Geosciences & Remote Sensing Symp. (IGARSS '83), San Francisco, 31 August - 2 September 1983.
Katsaros, K.B., Taylor, P.K., Alishouse, J.C. & Lipes, R.G. (1981) Quality of Seasat SMMR (Scanning Multichannel Microwave Radiometer) atmospheric water determinations, in: Oceanography from Space (ed. J.F.R. Gower), Plenum, 691-706.
Kaufeld, L. (1981) The development of a new Beaufort equivalent scale, Meteorol. Rundsch, 34, 17-23.
Large, W.G. & Pond, S. (1981) Open ocean momentum flux measurements in moderate to strong winds, J. Phys. Oceanog. 11, 324-336.
Large, W.G. & Pond, S. (1982) Sensible and latent heat flux measurements over the ocean. J. Phys. Oceanogr. 12, 464-482.
Lipes, R.G. (1982) Description of Seasat Radiometer status and results, J. Geophys. Res. 87, 3385-3398.
Liu, W.T., K.B. Katsaros and J.A. Businger (1979) Bulk parameterizations of air-sea exchanges of energy and water vapour including the molecular constraints at the interface. J. Atmos. Sci. 36, 1722-1735.
Liu, W.T. (1983) Remote sensing of latent heat flux (this volume).
Lovejoy, S and Austin, G.L. (1980) The estimation of rain from satellite borne radiometers. Quart. J. Roy. Meteorol. Soc. 106, 255-276.
McAlister, E.D., McLeish, W. (1970) A radiometric system for air-borne measurement of the total heat flow from the sea, Appl. Optics 9, 2697-2705.
McAlister, E.D., W. McLeish and A. Corduan (1971) Airborne measurements of the total heat flux from the sea during

BOMEX. J. Geophys. Res. 76, 4172-4180.
Macklin, S.A. and Guymer, T.H. (1980) Inter-platform comparisons of JASIN WMO observations, Jasin News (15), 5-9 (Unpublished manuscript).
McMurdie, L.A. (1983) Seasat Scanning Multichannel Microwave Radiometer (SMMR) observation of North Pacific Cyclones. M.Sc. Thesis, Dept. of Atmospheric Sciences, Univ. of Washington, Seattle.
Minnett, P.J., Zadovy, A.M. and Llewellyn-Jones, D.T. (1983) Satellite measurements of sea surface temperature for climate research (this volume).
Niiler, P.P. & Richardson, W.S. (1973) Seasonal variability of the Florida Current. J. Mar. Res. 31, 144-167.
Pierson, W.J. (1978) The potential impact of Seasat-A on the study of the planetary boundary layer over the ocean, in: Turbulent fluxes through the sea surface, wave dynamics, and prediction (Favre, A. & Hasselmann, K. eds.), Plenum, 507-530.
Pollard, R.T., Guymer, T.H. & Taylor, P.K. (1983) Summary of the JASIN 1978 field experiment, Phil. Trans. Roy. Soc. Lond., A308, 221-230.
Pond, S. Large, W.G., Miyake, M., and Burling R.W. (1979) A Gill twin propeller-vane anemometer for flux measurements during moderate and strong winds, Bound. Layer Meteorol. 16, 351-364.
Quayle, R.G. (1980) Climatic comparisons of estimated and measured winds from ships, J. Appl. Meteorol. 19, 142-156.
RAL (1982a) Workshop on next generation operational satellite meteorological sounding systems, Conf. report (Abingdon 2-5 November 1982), Rutherford Appleton Laboratory, Didcot, UK.
RAL (1982b) The Along Track Scanning Radiometer with Microwave Sounder ATSR/M, Instrument proposal, Rutherford Appleton Laboratory, Didcot, UK.
Reed, R.K., and Elliott, W.P. (1979) New precipitation maps for the North Atlantic and North Pacific Oceans, J. Geophys. Res., 84, 7839-7846.
Reitan, C.H. (1963) Surface dew point and water vapour aloft, J. appl. Met. 2, 776-779.
Riehl, H. (1979) Climate and weather in the tropics, Academic Press, London, 611 pp.
Robinson, G.D. (1966) Another look at some problems of the air-sea interface, Quart. J. Roy. Meteorol. Soc. 92, 451-465.
Robinson, I.S., Wells, N.C. and Charnock, H. (1983) The sea surface thermal boundary layer and its relevance to the measurement of sea surface temperature by airborne and spaceborne radiometers, Report: Contract F60/G6/06, Dept. of Oceanography, Univ. of Southampton.
Robson, M.E. (1973) Deficiencies in shipboard observations, in: WMO (1973), 76-81.
Ruprecht, E. and Gray, W.M. (1976) Analysis of satellite-observed

tropical cloud clusters. II. Thermal, moisture and precipitation fields, Tellus 28, 414-426.

Ruprecht, E. Medrow, W. and Raschke, E. (1983) Comparison of rainfall rates derived from radar and Nimbus 5 microwave observations in the tropical Atlantic, in: Street-Perrot et al. (1983), 141-146.

Ruttenberg, S. (1981) Needs, opportunities and strategies for a longterm oceanic sciences satellite program , NCAR Tech. Note TN-185, Boulder, 72pp.

Salby, M.L. (1982a) Sampling Theory for Asynoptic Satellite Observations, I. Space-Time Spectra, Resolution, and Aliasing. J. Atmos. Sci. 39, 2577-2600.

Salby, M.L. (1982b) Sampling Theory for Asynoptic Satellite Observations, II. Fast Fourier Synoptic Mapping. J. Atmos. Sci. 39, 2601-2614.

Shearman, R.J. (1983) The Meteorological Office main marine data bank, Met. Mag. 112, 1-10.

Smith, S.D., Katsaros, K.B. and Oost, W.A. (1983) HEXOS-Humidity Exchange Over the Sea: Scientific Plan, Canadian Techn. Rep. Hydrog. & Ocean Sci. (21), 47 pp.

Street-Perrott, A., Beran, M. and Ratcliffe, R. eds (1983) Variations in the Global Water Budget, D. Reidel, Dordrecht, 518 pp.

Strong A.E. [1982] Intercalibrations of directly-measured and remotely-sensed marine observations. Marine Meteorol. and Related Oceanographic Activities (9), 21 pp.

Stubbs, M.W. (1981) New code for reporting surface observations - an introduction, Weather, 36, 357-366.

Taylor, P.K. (1982) Remote Sensing of atmospheric water content and sea surface latent heat flux, Proc. Conf. Remote Sensing and the Atmosphere, Remote Sensing Soc., Univ. of Reading, 265-272.

Taylor, P.K., Guymer, T.H., Katsaros, K.B. & Lipes, R.G. (1982) Atmospheric water distributions determined by the Seasat Multichannel Microwave Radiometer, in Street-Perrott et al. (1983), 93-106.

Taylor, P.K. (1983a) The scanning multichannel microwave radiometer - an assessment, in: Satellite Microwave Remote Sensing (T.D. Allan ed.), Ellis Horwood Ltd., Chichester, 526 pp.

Taylor, P.K. (1983b) Variability of atmospheric water content and the remote sensing of latent heat flux. I.U.G.G. Interdisciplinary Symposia, Abstracts vol. 2, 670.

Tomasi, C. (1981) Determination of the total precipitable water by varying the intercept in Reitan's relationship, J. appl. Met. 20, 1058-1069.

Tucker, G.B. (1961) Precipitation over the North Atlantic Ocean, Quart. J. Roy. Meteorol. Soc. 87, 147-158.

Verploegh, G. (1956) The equivalent velocities for the Beaufort estimates of the wind force at sea, Mededelingen en

Verhandelingen (66), K.N.M.I., 38 pp.

Verploegh, G. (1967) Observation and analysis of the surface wind over the ocean, Mededelingen en Verhandelingen (89), K.N.M.I., 67 pp.

Vockeroth, R.E. and Cook J.M. (1973) A proposed shipboard meteorological data system, in WMO (1973), 426-437.

WMO (1962) Precipitation measurements at sea, C.M.M. Tech. Note 47, (WMO-No. 124 TP55), 18 pp.

WMO (1969) Oceanic part of the hydrological cycle, Reports on WMO/IHD, Projects, No. 11, 71 pp. + figs.

WMO (1970) The Beaufort scale of wind force (technical and operational aspects). Reports on Marine Science Affairs, No. 3, 22 pp.

WMO (1973) Means of acquisition and communication of ocean data, Vol. II Surface, sub-surface and upper-air observations, Reports on marine science affairs, No. 7, 600 pp.

WMO (1976) Oceanic Water Balance (WMO-No. 442), 112 pp.

WMO (1977) Guide to Marine Meteorological Services (WMO-No. 471).

WMO/ICSU (1981) Summary of Data Collected, Global Weather Experiment FGGE Operations Reports (a) Vol. 3: 27 November 78-6 March 79, (b) Vol. 4: 7 March-4 July 1979, (c) Vol. 5: 5 July-1 December 1979.

Weare, B.C. and P.T. Strub (1981) The significance of sampling biases on calculating monthly mean oceanic surface heat fluxes. Tellus 33, 211-224.

Weare, B.C., P.T. Strub and M.D. Samuel (1981) Annual mean surface heat fluxes in the tropical Pacific Ocean, J. Phys. Oceanogr. 11, 705-717.

Wentz, F.J., Cardone, V.J. and Fedor, L.S. (1982) Intercomparison of wind speeds inferred by the SASS, Altimeter, and SMMR.J. Geophys. Res. 87 (C5), 3378-3384.

THE SOLAR HEAT INPUT INTO THE UPPER MARINE ENVIRONMENT

N.K. Hojerslev

Institute of Physical Oceanography
Copenhagen, Denmark

ABSTRACT

A simple solar heat input model applicable to a stagnant, incompressible and horizontally stratified ocean is presented. The model demands only a single optical input parameter in order to work. Three (four) such different input parameters have been selected for the reason that they are routinely measured on a global scale ensuring maximum solar heat input informations. The three versions of the model compare favorably with other more elaborate models.

1. INTRODUCTION

The solar heat input into the upper marine environment as a function of water depth is dependent on a large variety of parameters which are: latitude, time of the year and of the day (solar elevation), optical properties of aerosols and hydrosols (light absorption and light scattering), cloud cover, the air mass and the pure seawater itself and finally the state of the sea surface. Some of these parameters are extremely difficult to measure either directly and/or simultaneously. For this reason it is here suggested to focus on a first order oceanic heat input model easy to extend.

The present model contains a minimum number of well-established optical parameters commonly measured either <u>in-situ</u> or remotely and is based on the following experimental results:

1) The downward quanta irradiance between 350-700 nm and

C. Gautier and M. Fieux (eds.), Large-Scale Oceanographic Experiments and Satellites, 247–256.

just below a flat sea surface can quite accurately be given in terms of quanta m^{-2} s^{-1} provided there is a cloud-free sky and a known solar elevation.

2) The ratio of the downward quanta irradiance and the downward energy irradiance both between 350-700 nm is almost constant for all water types from the sea surface and down to a depth above which 99% of the incoming visible daylight has been transformed into sensible heat.

3) For a standard atmosphere the total incoming downward energy irradiance (350-3000 nm) to the sea surface can be expressed in terms of the downward energy irradiance (350-700 nm) and consequently in terms of the downward quanta irradiance (350-7000 nm).

4) Any percentage level for the downward quanta irradiance (350-700 nm) can be given in terms of the depth of the euphotic zone for seemingly all oceanic and coastal water types. This depth is commonly measured on a worldwide basis either <u>in-situ</u> or remotely.

5) The depth of the euphotic zone is sufficient for calculating the almost total transformation of water penetrating daylight into sensible heat at various depths. This depth varies typically between 40m and 100m for oceanic waters and typically from 10m to 40m for coastal waters dependent on the water turbidity.

Within the upper one or two meter water column most of the infrared daylight is absorbed. Since there are quite a few models dealing with this particular aspect, it will not be treated at length in this context. Here, the emphasis has been clearly to describe the heat input into deeper water layers due to the daylight in order to demonstrate that this input is linked to the depth of the euphotic zone or alternatively to the turbidity of the water column.

2. THEORY

The equation for heat transfer in a Cartesian coordinate system with the z-axis oriented downwards writes in its general form:

$$\frac{\partial(\rho c_p T)}{\partial t} + \vec{v}\cdot\vec{\text{grad}}(\rho c_p T) + \rho c_p T \,\text{div}\,\vec{v} =$$

$$\frac{\partial}{\partial x}\left(K_x \frac{\partial(\rho c_p T)}{\partial x}\right) + \frac{\partial}{\partial y}\left(K_y \frac{\partial(\rho c_p T)}{\partial y}\right) + \frac{\partial}{\partial z}\left(K_z \frac{\partial(\rho c_p T)}{\partial z}\right) \qquad (1)$$

$$- \frac{\partial E_x}{\partial x} - \frac{\partial E_y}{\partial y} - \frac{\partial E_z}{\partial z}$$

Assuming a stagnant, incompressible and horizontally stratified ocean with respect to heat and hydrosols, Equation (1) reduces to:

$$\rho \, c_p \frac{\partial T}{\partial t} = - \frac{\partial E(z)}{\partial z} \qquad (2)$$

in which T is the temperature and E(Z) the so-called total vector irradiance defined as:

$$E(z) = \int_{\lambda=0}^{\infty} E(z,\lambda)\, d\lambda = \int_{\lambda=0}^{\infty} \int_{4\pi} L(z,\lambda) \cos\theta \, d\omega \qquad (3)$$

in which $L(z,\lambda)$ is the radiance at wavelength λ and $\int_{4\pi} d\omega$ stands for integration over all solid angles (for details see Hojerslev, 1978).

The vertical attenuation coefficient for the vector irradiance is defined as:

$$K(z,\lambda) = - \frac{1}{E(z,\lambda)} \frac{\partial E(z,\lambda)}{\partial z} \qquad (4)$$

so upon integration (beginning from just below the sea surface):

$$E(z,\lambda) = E(o,\lambda)\, e^{-\int_0^z K(z',\lambda)\, dz'} \qquad (5)$$

Accordingly, Equation (3) can be written as:

$$E(z) = \int_{\lambda=0}^{\infty} E(o,\lambda)\, d\lambda \; e^{-\int_0^z K(z',\lambda)\, dz'} \qquad (6)$$

Assuming all the $K(z,\lambda)$ coefficients independent of depth which is strictly not the case for even a homogeneous water mass (e.g. Hojerslev, 1980), Equation (6) becomes:

$$E(z) = \int_{\lambda=0}^{\infty} E(o,\lambda)\, e^{-K(z,\lambda)\cdot z}\, d\lambda \qquad (7)$$

and can be approximated into:

$$E(z) = E(0) \sum_{i=1}^{n} a_i e^{-k_i z} \qquad (8)$$

in which $\sum_{i=1}^{n} a_i = 1$. The number of $\underline{n}$ ranges from unity and upwards depending on the models applied. It should be noted that often Equation (7) is erroneously written as:

$$E(z) = \int_{\lambda=0}^{\infty} \int_{z=0}^{z=z} E(0,\lambda)\, e^{-k(z,\lambda)}\, d\lambda\, dz$$

(e.g. Zaneveld and Spinrad, 1980 or Simpson and Dickey, 1981). Moreover, if $\underline{n}$ is chosen too large the practical applicability of the model given by Equation (8) can be limited. A direct measurement of E(Z) instead would be more appropriate. It seems therefore natural to establish a model for E(Z) into which currently measured optical parameters enter.

3. THE MODEL

In case of monochromatic light, integration of the scalar equation of radiative transfer leads to:

$$\frac{\partial E(z,\lambda)}{\partial z} = \left(1 - R(z,\lambda) + \frac{1}{k_d(z,\lambda)} \frac{dR(z,\lambda)}{dz}\right) \frac{\partial E_d(z,\lambda)}{\partial z} \qquad (9)$$

in which $R(z,\lambda)$ is the irradiance ratio, $E_d(z,\lambda)$ the downward irradiance and $k_d(z,\lambda)$ the vertical attenuation coefficient for $E_d(z,\lambda)$

Since $\frac{1}{k_d}\frac{dR}{dz} \ll R$ and $R < 0.1$ for all wavelengths and all waters (Hojerslev, 1980) this term is assumed to be zero. Equation (9) is now integrated and approximated by:

$$E(z) = \int_{\lambda=0}^{700} (1 - R(\lambda)) E_d(z,\lambda)\, d\lambda + \int_{\lambda=700}^{3000} E_d(z,\lambda)\, d\lambda \simeq$$

$$\int_{\lambda=350}^{700} E_d(z,\lambda)\, d\lambda + \int_{\lambda=700}^{3000} E_d(z,\lambda)\, d\lambda \equiv E_1(z) + E_2(z) \qquad (10)$$

Note that the shift in the lower integration limit and putting $1-R(\lambda)=1$ counterbalance approximately one another. The ratio of the downward irradiance between 350-700 nm (the first term $E_1(Z)$ and the downward quanta irradiance in the same spectral range, $q(Z)$, can be assumed independent of depth within the euphotic zone for all water types (Jerlov, 1978). Moreover, the following two relations have been reported (e.g. Jerlov, 1974):

$$\frac{q(0)\ \text{quanta}\ m^{-2}s^{-1}}{E_1(0)\ \ W\ m^{-2}} = 2.75 \times 10^{8} \qquad (11)$$

(for white light the conversion factor amounts to $2.64.\ 10^{18}$).

$$E_1(0) \simeq 0.44\,(E_1(0) + E_2(0)) \qquad (12)$$

The vertical variation of $E_1(Z)$ can thus be given in terms of $q(Z)$. From Equations (10 and 12):

$$E(z) = E(0)\left[0.44\, e^{-\int_0^z K_1(z')dz'} + 0.56\, e^{-\int_0^z K_2(z')dz'}\right] \qquad (13)$$

in which

$$K_1(z) = -\frac{1}{q(z)}\frac{\partial q(z)}{\partial z}$$

$$K_2(z) = -\frac{1}{E_2(z)}\frac{\partial E_2(z)}{\partial z} \qquad (14)$$

In order to calculate $K_2(Z)$ it will be assumed that $K_2(Z)$ is independent of both depth and concentration of suspended matter (e.g. Jerlov, 1976 and Bethoux, 1977). Applying Equation (13) and the above assumptions for $K_2(Z)$ to Jerlov's optical water type I characterized by $q(Z= 1m_{-1})\%$, (Jerlov, 1978) and $E(Z= 1m)\%$, (Jerlov, 1976), $K_2 = 2.35\ m^{-1}$ and

$$E(z) = E(0)\left[0.44\, e^{-\int_0^z k_1(z',\lambda)\,dz'} + 0.56\, e^{-2.35 z}\right] \quad (14)$$

By combination of Equations (11) and (12), E(0) can be given in terms of q(0) and <u>vice versa</u>. If the sky is cloud-free

$$q(0) = 1200 \sin^{1.4} h_s \times 10^{18} \text{ quanta m}^{-2}\text{s}^{-1} \quad (15)$$

for all solar elevation ranges (Hojerslev, 1980). Direct measurements of E(0+), (Bethoux, 1977), subtracted the Fresnel reflection at the air-sea interface assuming no sky light, gives approximately E(0) which also demonstrates a $\sin^{1.4} h_s$ dependence valid for the solar elevation range $1° \leq h_s \leq 60°$. It seems therefore justified from Equations (11) and (12) to conclude that:

$$\frac{q(0) \text{ quanta m}^{-2}\text{s}^{-1}}{E(0)\ \text{W.m}^{-2}} = 12 \times 10^{18} \quad (16)$$

for the whole solar elevation range provided there are no clouds.

If clouds are present a correction term of the form $(1-0.70C^2)$ can be applied to Equation (14) in which C is the cloudiness expressed in fractions of tenths (Kondratyev, 1969). Equation (14) offers a variety of advantages:

a) $K_1(Z)$ is a pertinent parameter in photosynthetic studies and is thus routinely measured. It can be considered as the first and primary optical input parameter to the solar heat input model.

b) $K_1(Z)$ can be determined from NIMBUS-7, CZCS data or from rapid color measurements of the sea (Hojerslev, 1980 and 1982).

c) $K_1(Z)$ can be calculated from a simple monochromatic irradiance measurement around 500 nm (Hojerslev, 1981).

d) $K_1(Z)$ is one of the key parameters in Jerlov's optical classifications of water masses (Jerlov, 1976 and 1978).

e) $K_1(Z)$ is almost independent of the surface light conditions, i.e. also independent of solar elevation (Hojerslev, 1974a, 1974b and 1980) in different water masses. This implies that $\frac{\partial T}{\partial t}$ in Equation (2) is solely determined by E(0) - or alternatively by q(0) - multiplied by the term $(1 - 0.70C^2)$.

f) The Secchi disk depth equals approximately $z_q(10\%)$, (see Equation (21) and Hojerslev, 1978 and 1982).

In the following, the downward quanta irradiance is considered as a single input parameter to be used in the model and E(0) is equalled to unity for reasons of simplicity. The depths of the 50%, 30%, 10%, 3% and 1% levels for the downward quanta irradiance are mutually related to one another through relations which were observed in the course of a year on Fladen Ground (Hojerslev, 1982):

$$z_q(50\%) = 0.20\ z_q(10\%) \tag{17}$$

$$z_q(30\%) = 0.48\ z_q(10\%) \tag{18}$$

$$z_q(3\%) = 1.58\ z_q(10\%) \tag{19}$$

$$z_q(1\%) = 2.16\ z_q(10\%) \tag{20}$$

This is in good agreement with other findings (i.e. Aas, 1980, Hojerslev, 1978 and Jerlov, 1977). The first term in Equation (14) can be expressed in terms of $z_q(10\%)$ by:

$$\int_0^{z_q(10\%)} K(z')\,dz' = \ln 10 \equiv \langle K \rangle\, z_q(10\%) \tag{21}$$

Consequently, if $z_q(10\%)$ is determined through a <u>single</u> measurement then application of Equations (17)-(21) to Equation (14) gives E(Z) in a straightforward manner and the model serves its purpose by calculating the first derivative of E(Z) with respect to depth - see Equation (2).

4. SUMMARY

In order to test the validity of the optical model, E(Z), for the selected optical water types I, IB, II, III and 3 (Jerlov, 1976), are compared with the values predicted by the model. The input parameter will be $z_q(10\%)$ corresponding to the above-mentioned water types. The results are depicted in

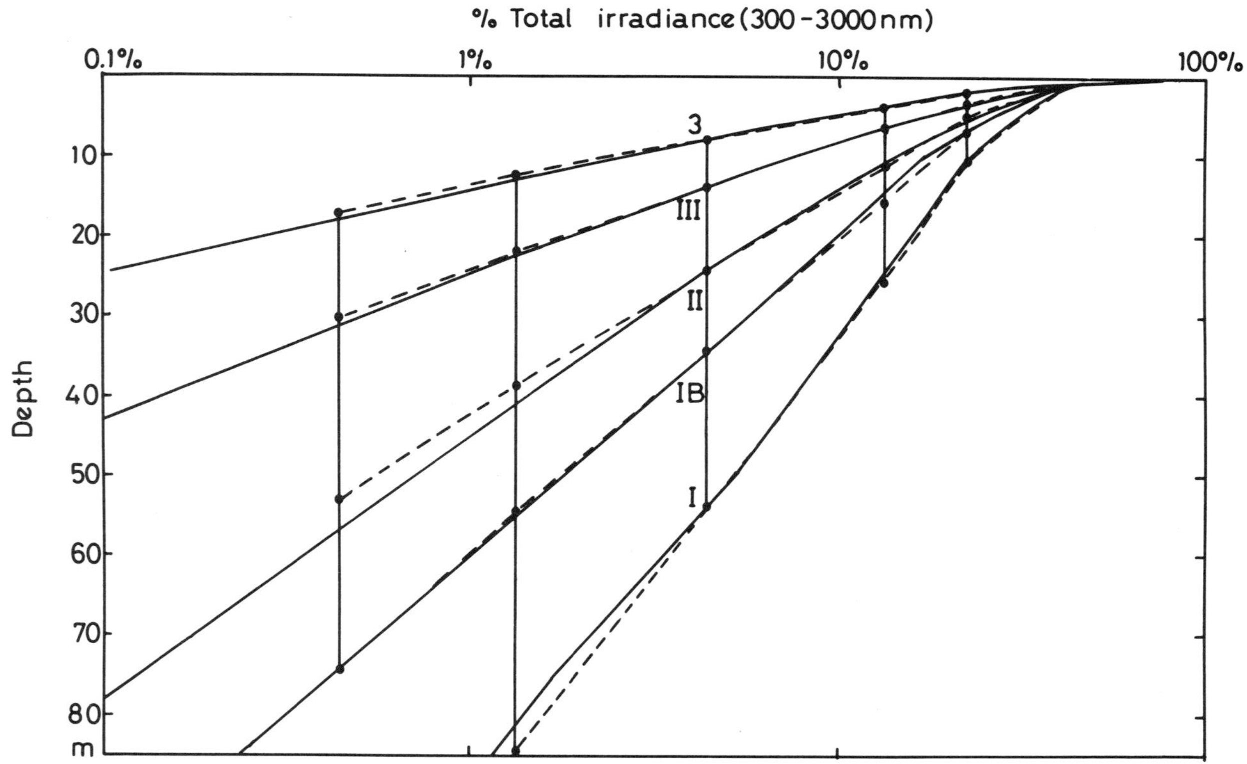

Fig. 1 Depth profiles for the total downward irradiance in percent of surface irradiance for the optical water types I, IB, II, III and 3. Solid curves are from Jerlov (1976). Dashed curves are as computed from the model applying $z_q(10\%)$ as the only optical input parameter.

Figure 1 and they seem satisfactory when compared to other models (e.g. Zaneveld and Spinrad, 1980 or Simpson and Dickey, 1981).

As mentioned earlier, $z_q(10\%)$ can be derived from the water leaving Nadir radiance at 443 nm and 520 nm (e.g. Hojerslev, 1980 and 1982) so the NIMBUS-7, CZCS, Coastal Zone Color Scanner is one obvious tool in modelling the heat input to the marine environment.

It is also feasible to determine q(Z), or alternatively $z_q(10\%)$, in terms of vertical profiles of the downward irradiance around 500 nm with a sufficient accuracy (Hojerslev, 1981). Such a monochromatic daylight measurement is quite independent of the state of the sea among other favorable aspects (e.g. Hojerslev, 1978 for further details).

A Secchi disk measurement gives approximately $z_q(10\%)$ but the state of the sea is crucial. A wave height (2X amplitude) of 1m along the Secchi disk string reduces the Secchi disk depth by 40% (Hojerslev, 1982).

ACKNOWLEDGEMENTS

The author was invited to participate at the NATO ARW meeting in Corsica in the fall, 1983. I would like to express my sincere thanks to the organizers who made my participation possible.

REFERENCES

Aas, E., 1980, "Relations between Total Quanta, Blue Irradiance and Secchi Disc Observations in the Norwegian and Barents Seas", Rep. Inst. Phys. Oceanogr., Univ. Copenhagen, 42, 11-27.

Bethoux, J.P., 1977, "Contribution à l'Etude Thermique de la Mer Mediterranée", Thèse de doctorat d'Etat es Sciences Physiques. No d'Enregistrement au C.N.R.S., A.O., 24 Mars, pp. 192.

Hojerslev, N.K., 1974a, "Inherent and Apparent Optical Properties of the Baltic", Rep. Inst. Phys. Oceanogr., Univ. Copenhagen, 23, pp. 70.

Hojerslev, N.K., 1974b, "Daylight Measurements for Photosynthetic Studies in the Western Mediterranean", Rep. Inst. Phys. Oceanogr., Univ. Copenhagen, 26, pp. 38.

Hojerslev, N.K., 1978, Daylight Measurements Appropriate for Photosynthetic Studies in Natural Sea Water, J. Cons. int. Explor. Mer. 38, 131-146.

Hojerslev, N.K., 1980, Water Color and its Relation to Primary Production, Bound. Layer Met. 18, 203-220.

Hojerslev, N.K., 1981, "Optical Properties of Ocean Waters". In IAMAP Symposium on the Radiation Transfer in the Oceans and Remote Sensing of Ocean Properties. IAMAP Assembly, Hamburg 1981.

Hojerslev, N.K., 1982, Bio-optical Properties of the Fladen Ground: "Meteor" - FLEX-75 and FLEX-76, J. Cons. int. Explor. Mer. 40, 272-290.

Jerlov, N.G., 1974, "A Simple Method for Measuring Quanta Irradiance in the Ocean", Rep. Inst. Phys. Oceanogr., Univ. Copenhagen, 24, 7.

Jerlov, N.G., 1976, "Marine Optics", Elsevier Oceanography Series, 14. Elsevier Amsterdam, 2nd ed., 203.

Jerlov, N.G., 1977, Classification of Seawater in Terms of Quanta Irradiance, J. Cons. int. Explor. Mer. 37, 281-287.

Jerlov, N.G., 1978, "The Optical Classification of Sea Water in the Euphotic Zone", Rep. Inst. Phys. Oceanogr., Univ. Copenhagen, 36, 46.

Kondratyev, K.Y., 1969, "Radiation in the Atmosphere", International Geophysics Series, 12. Academic Press New York and London, pp. 912.

Simpson, J.J. and T.D. Dickey, 1981, Alternative Parameterizations of the Downward Irradiance and Their Dynamical Significance, J. Phys. Oceanogr. 11, 876-882.

Zaneveld, J.R.V. and R.W. Spinrad, 1980, An Arc Tangent Model of Irradiance in the Sea, J. Geophys. Res. 85, C9, 4919-4922.

SUMMARY OF DISCUSSION AND RECOMMENDATIONS ON ALTIMETRY

Moderator and Rapporteur:
J.F. Minster

The main purposes of the discussion were to propose a list of homework to be done in the near future using mainly the Seasat data, and to assess scientific consequences of the simultaneous flights of TOPEX, ERS-1 and NROSS.

I - Is the understanding of errors related to Seasat altimeter data adequate?
This includes:

1) Spectra for residual errors after corrections for ionospheric effects, tropospheric effects (humidity and atmospheric pressure), wave height, tidal models...

2) Spectrum for orbit error in particular its intermediate time scale components (a few days).

a) In particular the question has been raised of how these errors affect the surface height variability which can be deduced from Seasat data and its space-time spectrum.

b) It has been mentioned that considerable reduction of orbit errors will be obtained by adequate modelling of the causes for these errors. Reducing the latter necessitates an extended coverage for orbit tracking.

II - Comparison of Seasat data with other measurements.

1) Mesoscale variability as deduced from Seasat data is an Eulerian sea surface height variability. Buoys and current-meter historical data are respectively Lagragian and

C. Gautier and M. Fieux (eds.), Large-Scale Oceanographic Experiments and Satellites, 257–260.

Eulerian velocity measurements. Comparison of both requires:

a) to understand which part of the latter corresponds to geostrophic velocity.

b) to understand how Eulerian and Lagragian measurements are related.

These points can be in particular studied through eddy resolving models.

2) The comparison of the variability maps as deduced from altimetric data with the instantaneous flow visualization as observed from IR and visible satellite images needs to be studied. The requirements of such a study in terms of the image data set to be processed are to be defined.

3) Spectrum of variability needs to be studied not only along tracks but also along East West directions, between neighbouring tracks. Study of the spatial and temporal coherency of this variability will be useful for understanding its significance and eventually for following its evolution along the mean currents.

4) Extraction of tidal mode M2 from Seasat data has been very encouraging. It is suggested that adjustment of the models using satellite data can be made in a more direct way and will eventually be feasible for modes other than M2.

5) Evaluation of mean sea surface and its variability is an essential aspect for WOCE and TOGA. Further studies from Seasat data are necessary in particular with regards to the zonal variability in tropical regions. Work on simultaneous improvement of geoids and mean sea surface still needs to be done. In particular, adequate understanding of how altimeter data can improve the geoids to be subtracted from altimetric surface needs further analysis. For short distance scales, the extent to which bathymetric data can be used to deduce useful local geoids needs to be examined.

III - Some work in preparing future missions.

1) Understanding is needed of the long term changes in the corrections to be applied to give mean sea surface. This concerns in particular monitoring the long term deviations of the instruments, seasonal evolution of the orbits, seasonal characters of the ionospheric and tropospheric effects and semiannual tides.

2) The extent to which altimeter data can be used to estimate the seasonal heat storage of the surface ocean needs to be examined.

3) Models for tides need to be developed further in any case. Present models apparently estimate the ratio of the amplitude of different modes of similar period better than their actual amplitude. The extent to which this is true and can be used needs to be examined.

4) In principle, both altimeters and microwave radiometers identify heavy rain areas and the corresponding data are eliminated. Whether this can be used to quantitatively estimate the importance of heavy rains over the ocean needs to be studied. It is in particular necessary to estimate the occurence of heavy rain in areas where scatterometer wind measurements are made.

IV - Importance of simultaneous flight of TOPEX, ERS-1 and NROSS.

The simultaneity of the flights of TOPEX, ERS-1 and NROSS have implications at different levels.

1) To understand the dynamic response of the ocean to changes in surface wind forcing, simultaneous measurements of surface topography and windstress are required, especially in the tropics. This is one of the main objectives of TOGA. On the global scale, the most urgent task to be studied by WOCE, is to provide a data set for testing general circulation models for which simultaneity clearly is a major advantage, although some of the uncertainties in present models can be addressed with non simultaneous data.

2) Altimetric measurements at the precision expected during the TOPEX mission can be considered a new technique. It is desirable to cross compare such measurements in order to gain confidence on their reliability. Intercomparison of the instruments will benefit the three missions. In particular knowledge of the orbits of ERS-1 and NROSS will be greatly improved by cross comparison with that of TOPEX. This in turn might allow an improvement of TOPEX data through redundancy, cross calibration of the instruments and the data treatment and estimations outside of the domain of coverage of TOPEX. This has implications to be studied concerning the choice of the orbits for ERS-1 and NROSS and the use of calibration ground-based instruments (laser stations, pressure gauges). This concerns both the altimeters and the microwave radiometers.

3) The design analysis for an individual altimeter mission indicates that with appropriate compromises, it is possible to approach an unambiguous interpretation of the data in terms of the WOCE and TOGA objectives. However, these inevitable compromises have little margin for unexpected difficulties. The availability of a second simultaneous altimeter satellite in a

different orbit allows a more flexible approach, yielding greater confidence that the scientific objectives can unambiguously be achieved. These benefits include extended areal coverage, higher sampling density, dealiasing of the tides and system intercomparison.

4) All missions require extensive ground based measurements. It will represent a major effort for all the oceanographic communities. The planning for this effort should take into account simultaneously all aspects of the three space systems.

SUMMARY OF DISCUSSION AND RECOMMENDATIONS ON SEA-SURFACE TEMPERATURE

Moderators and Rapporteurs:
P. Deschamps & F. Webster

Sea-surface temperature (SST) is required by the TOGA and WOCE experiments for several purposes; the primary need is as an atmospheric boundary condition; for example, changes in SST as small as 0.5K can shift atmospheric convection patterns, particularly in the tropics.

SST is also needed for heat-flux calculations. The net heat flux through the ocean-atmosphere interface is usually calculated as the sum of net shortwave + net longwave + latent + sensible heat fluxes. The last three terms depend upon the sea-surface temperature. At wind speed above 5 m/s an error in SST of about 0.5K leads to a total error of more than 10 W/m^2. For the latent and sensible heat fluxes the relevant quantity is the air and sea temperature difference.

Monthly SST anomalies have been heavily used in describing the El Nino/Southern Oscillation phenomenon. Anomalies of 0.5K are significant, and thus an accuracy better than that currently available is required. However, one can begin to describe the large amplitude signals using the most accurate existing instruments, in favorable conditions.

The Workshop reviewed the current technology. Reports were given on results from the Advanced Very High Resolution Radiometer (AVHRR) instrument in the infrared region onboard NOAA satellites, and the Scanning Multi-spectral Microwave Radiometer (SMMR) experiment in the microwave region onboard the Seasat and NIMBUS-7 satellites. The Along-Track Scanning Radiometer (ATSR) experiment in the infrared region, to be

C. Gautier and M. Fieux (eds.), Large-Scale Oceanographic Experiments and Satellites, 261–263.

launched in 1988 on ERS-1, promises to provide improved atmospheric corrections that will lead to a better accuracy of SST measurement.

BULK OR SKIN TEMPERATURE?

The kind of surface temperature that should be estimated from satellite observations is the subject of some controversy. The "skin" temperature, just at the surface, is the temperature that is instantaneously seen by the radiometer. For month-long averages, however, it is more likely that the "bulk" temperature, that is the temperature of the sea below the diurnal thermocline (say about one meter depth) is the important quantity for the atmospheric boundary condition. Bucket temperatures, often taken to be a standard for shipboard measurement, do not consistently sample in a way that can be easily related to either of these quantities, and their use presents many problems. Buoy surface-temperature measurements, similarly, have an uncertain interpretation.

Because of their differing uses, both skin and bulk temperatures need to be collected. For climate studies, algorithms using the SST could be based on either skin or bulk surface temperature; for purposes of the WCRP, it seems preferable to encourage the use of bulk temperature averages as the more stable and reliable quantity. Algorithms in turn should be tailored to the use of bulk temperatures. This requires further work to relate the skin temperature to the bulk temperature, with studies of the validity of parameterizing the difference between them. That is, can we correct the skin temperature measured from space so that it can be used in bulk formulae? Better still, would be to devise new expressions to make direct use of the skin temperature.

ACCURACY LIMITS

Existing techniques for SST measurement from satellite seem to have a limit of about 0.6K rms error. Efforts to reduce this "error" have been made, but it has proven difficult. There are a number of difficulties that arise when comparing satellite observation with surface measurements. There is generally a difference in position and time of the measurements, and the wide field of view of instruments such as the SMMR makes comparison with point measurements questionable. Satellites sense the skin of the ocean, while surface observations are collected at a depth that varies from a few centimeters to several meters. Subpixel cloud contamination of the infrared observations under partly cloudy conditions also accounts for error.

As a "homework" assignment before proceeding further, it is becoming urgent to carry out studies to examine the nature of satellite SST measurement errors:

- Are the errors random or systematic?
- Are the errors geographically and seasonally variable?
- Are the errors due to:
    - data-handling algorithms?
    - satellite instrumental errors?
    - ground-truth errors?
    - natural atmospheric or oceanic variability?

RECOMMENDATIONS

A number of steps should be taken next:

1. The scientific need to improve satellite SST accuracy should be defined. An operational improvement on current levels (about 0.5K) would likely be expensive, and the need should be defined in advance in order to justify the effort.

2. A standard for identifying SST "errors" in terms of oceanic and atmospheric conditions should be developed. This is likely to be regionally variable.

3. A design study should also be carried out to define satellite-measured SST accuracy standards.

4. The regional variability of cloud screening effects should be examined, with particular attention on the tropics.

5. The relation between skin and bulk temperatures should be further examined, and quantitative means for relating them should be developed.

6. More attention should be given to accurate radiometric ground truth measurements. With present ground truth data it remains an open question whether or not satellite measurements of SST have reached an imposed limit of accuracy.

SUMMARY OF DISCUSSION AND RECOMMENDATIONS ON SURFACE WIND VELOCITY AND STRESS

Moderator and Rapporteur:
J. McWilliams

The following is a summary of several presentations and discussions on surface wind and stress measurements in the context of the WOCE and TOGA experiments, although the conclusions are likely to be apt in a broader context as well. It is important that these conclusions be conveyed to the appropriate organizers and sponsors of these experiments for their possible implementation.

A primary conclusion is that the stress itself, not some low-level wind speed, is the quantity of greater utility. The vector stress is by definition the momentum transfer to the ocean, and the scalar stress is a fundamental scaling parameter of both the sensible and latent heat fluxes. This implies that empirical regression algorithms for satellite "wind" measurements (e.g., scatterometry, altimetry) should be targeted to stress, not wind, and that data processing algorithms, such as averaging or smoothing, should be applied directly to stress values, either in a vector or scalar sense depending upon the intended use. If, alternatively, the satellite measurements are processed to yield winds directly, then the subsequent conversions to momentum and heat fluxes using bulk formulae will contain additional errors due to this conversion.

For similar reasons the calibration standard in developing the satellite algorithms should be direct measurements of the fluxes rather than bulk parameterizations. Furthermore, efforts should be made to obtain the calibration measurements with a degree of areal averaging, so that sampling errors are diminished in the comparison with the remote measurements.

C. Gautier and M. Fieux (eds.), Large-Scale Oceanographic Experiments and Satellites, 265–268.

These considerations are particularly important insofar as the satellite data are potentially at least as accurate as the parameterization estimates, particularly when averaged to the space and time scales of interest to WOCE and TOGA.

The present algorithm for the Seasat scatterometer has been shown to have dependences on sea surface temperature and atmospheric stability, and it is possible that there are other dependences as well. These are examples of bias rather than randomness in scatterometer errors. It is important to make additional calibration studies, over a wide range of geographical circumstances, to expose these biases and, if possible, correct them in the algorithm. If, as appears likely, scatterometers are deployed which measure in different wave bands, then the different instruments must each be calibrated, and it would be useful to arrange some circumstances where they can be simultaneously calibrated. For example, pre-launch aircraft flights over in situ arrays could be done.

The scatterometer algorithms are presently empirical. Obviously it would be of great value to improve theoretical understanding of the relationship between surface stress and the short surface waves which affect the backscatter, either to provide a theoretically derived algorithm or to check for possible biases which might be in the empirical algorithms. This theoretical problem is quite difficult, and no breakthrough can be expected prior to WOCE and TOGA. Nevertheless, theoretical studies should be pursued as an investment in the future.

For WOCE and TOGA, scatterometry measurements will still be sufficiently uncertain and unfamiliar that a substantial program of comparison measurements would be of great value. These measurements are of two types: high quality surface wind measurements from ships and buoys and sea-level pressure measurements from buoys. Both of these involve parameterizations to yield stress estimates, and both will require substantial augmentation of existing measurement programs. Ship winds will inevitably be limited mostly to commercial shipping routes, buoy winds might be sought away from these routes, and buoy pressures could be obtained globally. An additional argument for these in situ programs is as protection against possible failures of the satellite instruments during WOCE and TOGA. It is, however, a reasonable expectation that satellite measurements by themselves eventually will constitute an adequate observing system for stress.

Studies are needed of the sensitivity of oceanic response to errors in surface stress in a variety of physical models. This should be the basis for determining the accuracy requirements in WOCE and TOGA. Many different types of models should be used since many different processes are relevent to these experiments, and no existing model is close to comprehensive in its processes. Examples of model types are forced equatorial waves, Ekman transports, Sverdrup transports, mixed-layer evolution, forced mesoscale eddies, and Equatorial Undercurrent and Antarctic Circumpolar Current transports.

A design study is needed for the optimal combination of orbits for the two anticipated scatterometers in WOCE and TOGA. It should also address the utility of additional scalar stress measurements from the several altimeters to be flying at the same time and whether adjustments in their orbits would be warranted. Consideration should be given to assuring sufficient in flight intercalibration between the different instruments, and contingency plans should be made for the possibility of instrument failures.

A technical issue in obtaining satellite stress measurements is the necessity for discarding samples with excessive rainfall in view. A study is needed of whether the instruments planned for WOCE and TOGA have adequate indicators of rainfall and whether the sampling biases associated with discarding winds in rain are tolerable. The answer to the latter question is more likely to be favorable for vector stress than scalar.

The possibility should be explored that passing the stress measurements through a meteorological assimilation and forecast model, together with all of the other observations required for such a model, would yield an improved analysis of the global stress field compared to more direct analyses of the measurements. Obviously, the success of such a procedure is dependent upon the ability of the models to assimilate surface stress with sufficient weighting and the accuracy of their surface physics. Both of these issues are presently in doubt, and model developments would be needed to realize this possibility. This approach could also be taken with other surface fluxes.

TOGA will begin before any new scatterometers are flown. It is possible that scalar stress measurements from the Geosat altimeter may be useful in the interim. An inquiry should be made as to the public availability of these measurements.

The preceding remarks indicate a collection of tasks which should be accomplished in preparation for WOCE and TOGA. This collection is undoubtedly incomplete since it was created in limited discussions. We believe it is important that ways be found both to continue the discussions of necessary tasks and to arrange for their accomplishment.

RECOMMENDATIONS

1. To use the wind stress and not the wind velocity.

2. To use direct measurements of stress as calibrations standards.

3. To make calibration studies to investigate the scatterometer biases.

4. To improve theoretical understanding between surface stress and short surface waves.

5. To implement an intercomparison measurement program between satellite measurements and ship and buoys wind and pressure measurements.

6. To perform model sensitivity studies of oceanic response to surface stress errors.

7. To assess optimal orbit combination of the two anticipated scatterometers.

8. To ensure adequate data flaging for rainfall in particular.

9. To explore possibilities of surface stress assimilation in meteorological models and producing surface stress as a model analysis.

10. To inquire about Geosat stress (scalar) data availability.

SUMMARY OF DISCUSSIONS ON OCEAN SURFACE HEAT FLUX

Moderator and Rapporteur:
W.G. Large

Ultimately global climatology and variability of the net surface heat flux, Q, will rely heavily on satellites. One approach is to estimate the various components, SW↑↓ (net solar), LW↑↓ (net longwave), H_L (latent) and H_s (sensible), separately then add them together. Because the components have different signs, the magnitude of the annual Q, as estimated from bulk parameterizations of each of its four components, over most of the world's oceans is less than 40 W/m^2 and is thus usually less than the cumulative uncertainty in estimating it. Therefore, in order to develop and improve satellite techniques to the point where the error in Q is at least smaller than its local mean over most of the ocean, these techniques need to be calibrated and verified against direct surface measurements and not against present parameterization schemes. An additional problem, based on the experience with ship data, is that the observations from a single polar orbiting satellite will not always be sufficient to keep the sampling error small.

Since the degree of success of the above approach is not certain, the inferential techniques, as explored for CAGE, are still attractive. Of most relevance to satellites is the calculation of Q as a residual of the Earth's radiation budget and the atmospheric divergence of heat flux. There is also the possibility of obtaining the non-solar part of Q in the upper ocean from near surface vertical temperature gradient methods, for example. Over some large ocean regions during certain times of year, changes in the heat storage may be closely related to Q. Ultimately any successful approach may require a combination of satellite and surface observations.

C. Gautier and M. Fieux (eds.), Large-Scale Oceanographic Experiments and Satellites, 269–272.

Discussions on each of the heat flux components revealed that at present satellite SW$\uparrow\downarrow$ techniques are far more promising than the others, with a satisfactory end in sight (see conclusions). Research into the others is still at a very basic level, and will be hampered by the difficulty of obtaining the necessary direct surface calibration measurements.

The net longwave flux at the surface is sensitive to profiles of temperature and humidity, and to cloud amount, optical thickness and base height. Satellites cannot at present sense all of these quantities, and cannot even foresee sensing cloud base directly. There is a possibility, however, that bulk formulae of the type LW$\uparrow\downarrow$ = f (SST, low level humidity, Cloud Cover) may be valid in a statistical sense if the variability of cloud and upper atmospheric variables vary rapidly and reasonably stationarily, so that a monthly averaged LW$\uparrow\downarrow$, depending only on the above satellite derivable variables, may be valid.

There is not yet any direct way of determining H_S and H_L by satellite. Methods of indirect H_L estimations to within ± 40 W/m^2 (monthly mean) have been suggested, but this has not been demonstrated globally and it may not be possible to improve the accuracy very much. In this situation the <u>in situ</u> surface observations must be regarded as important in providing flux measurements in their own right. Unfortunately, at present they are deficient both in accuracy and distribution. Accurate satellite observations of surface winds (discussed elsewhere) will be an important first step towards solving these H_S and H_L problems.

CONCLUSIONS AND RECOMMENDATIONS

1. A broad range of continuous research efforts are necessary before global surface heat flux estimates can be expected to achieve an accuracy better than Q itself. As motivation, large scale experiments planned for the next decade must establish their heat flux measurement requirements at the earliest possible date. These experiments also ought to assess the impact of these requirements not being fully satisfied. Some experiments may need to be restricted to regions where Q is large or to phenomena where Q is not important.

2. The pursuit of Q, other than as a sum of components, is very attractive and would be greatly aided if:

   i) Heat flux, total and components, became a standard diagnostic of atmospheric models.

ii) Residual calculations of Q could be reduced to an ocean basin or smaller region and verified.

iii) Direct vertical ocean heat flux (non-solar) measurement techniques (e.g., following the two-wave length radiometer method) were perfected.

3. Requirements for operational SW satellite coverage are:

i) Extension to high latitudes using polar orbiting satellites.

ii) Calibration with direct observations in a variety of geographical locations and under a full range of meteorological conditions. A study needs to be performed to determine how to sample the satellite pixel at the surface.

iii) A study is needed to determine the minimum necessary satellite sampling required to completely specify the SW field under study.

4. Progress towards satellite LW measurements requires:

i) A look at cloud climatology in various regions to check variability.

ii) Long calibration experiments, that must include the tropics, with direct surface LW measured to 5% to compare with bulk formulae that depend only on SST, near surface humidity, cloud amount and perhaps other satellite derivable parameters.

5. Actions that would improve the H_S and H_L situations include:

i) Improving the accuracy and coverage of the present surface *in situ* data set. The use of buoys in data sparse regions and the use of all the observations from ships would increase the coverage, but be of marginal benefit if the accuracy is not sufficient.

ii) Pursuing the potentially great improvement of satellite indirect observations as offered by profile information (e.g., TOVS, VAS, and eventually AMSU).

iii) Evaluating the incorporation of boundary layer models into indirect satellite methods. Experiments such as FIRE should have a component to investigate the relationship between heat fluxes and satellite parameters.

iv) Making available to scientific users (in suitable form), the data from SSMI and the wind data from the Geosat altimeter.

SUMMARY OF DISCUSSION AND RECOMMENDATIONS ON PRECIPITATION

Rapporteur:
C. Gautier

Although the Precipitation issue was not explicitly part of the meeting, the participants decided that a discussion on it was necessary for several reasons:

a) The knowledge of Precipitation is important for the salinity budget, an important parameter in the WOCE experiment.

b) An improved knowledge of Precipitation may have implications for some of the objectives of TOGA and WOCE.

c) Precipitation is an important output of atmospheric models.

d) Precipitation affects most of the oceanographic measurements from satellite and care must be taken in the analysis of this data to implement adequate rain flags.

Precipitation can be used as a tracer of water mass formation for WOCE. At present, we only have a general climatological picture of Precipitation and Evaporation over the globe. This knowledge has come from the analysis of a few in-situ gauge measurements, ship cloud reports and more recently from satellite observations. The Highly Reflective Cloud technique and microwave observations have given us a large-scale picture of the regions covered by Precipitation but only radar measurements have been transformed into area averaged rainfall rates. Thus over the ocean we have presently very few means of observations. New measurements techniques need to be investigated such as underwater accoustic noise measurements. This will require the transfer of existing standards to the open ocean.

C. Gautier and M. Fieux (eds.), Large-Scale Oceanographic Experiments and Satellites, 273–274.

RECOMMENDATIONS

As with the other processes discussed, we need:

1. to review the scientific objectives of TOGA and WOCE which require the knowledge of Precipitation (atmospheric heating rates, for instance for TOGA, or use as tracer for WOCE),

2. to characterize the scale and accuracy of both TOGA and WOCE requirements,

3. to quantify current uncertainties,

4. to investigate the nature of these uncertainties (random or systematic),

5. to establish calibration and measurements standards. It will be a very difficult task and therefore the necessity and precision of accomplishment needed for TOGA or WOCE must first be defined.

ACRONYMS AND DEFINITIONS

EXPERIMENTS AND PROGRAMS

FGGE	First GARP Global Experiment
GARP (GWE)	Global Atmospheric Research Program
GATE	GARP Atlantic Tropical Experiment
GWE (GARP)	Global Weather Experiment
HEXOS	Humidity Exchange over the Sea
JASIN	Joint Air-Sea Interaction Experiment
MILDEX	Mixed Layer Dynamics Experiment
MODE/ POLYMODE	Mid-Ocean Dynamic Experiment
MONEX	Monsoon Experiment
STREX	Storm Transfer Response Experiment
TOGA	Tropical Ocean and the Global Atmosphere
WCRP	World Climate Research Experiment
WOCE	World Ocean Circulation Experiment

GENERAL DEFINITIONS

GTS	Global Telecommunication System
OWS	Ocean Weather Station
VOF	Voluntary Observation Fleet

INTERNATIONAL BODIES

CCCO	Committee on Climatic Change and the Ocean
COSPAR	Committee on Space Research
ECMRWF	European Center for Medium Range Weather Forecast
ESA	European Space Agency
WMO	World Meteorological Organization

NATIONAL ORGANIZATIONS AND CENTERS

CNES	Centre National d'Etudes Spatiales (FRANCE)
FNOC	Fleet Numerical Ocean Center (USA)
JMA	Japanese Meteorological Agency (JAPAN)
NASA	National Aeronautics and Space Administration (USA)
NCAR	National Center for Atmospheric Research (USA)
NMC	National Meteorological Center (USA)

SATELLITES

DMSP	Defense Meteorological Satellite Program (USA)
ERB	Earth Radiation Budget (USA)
ERS-1	Earth Resource Satellite (EUROPE)
GEOS	Geodynamic and Experimental Oceanographic Satellite (USA)
Geosat	Geodynamic Satellite (USA)
GOES	Geostationary Operational and Environmental Satellite (USA)
GMS	Geostationary Meteorological Satellite (Japan)
Insat	Indian Geostationary Satellite (INDIA)
Landsat	Land Resource Satellite (USA)

Meteosat	European Geostationary Satellite (EUROPE)
MOSS-1 MOSS-2	Marine Observation Satellite System (Japan)
NIMBUS-7	Experimental Polar Orbiting Satellite (USA)
NOAA-7 NOAA-8	Operational Polar Orbiting Satellites (USA)
NROSS	Navy Remote Ocean Satellite System (USA)
Seasat	Experimental Ocean Satellite (USA)
TIROS	Television Infrared Operational Satellite
TOPEX	(Ocean) Topography Experiment (USA)
TOPEX/ POSEIDON	(Ocean) Topography Experiment (USA and FRANCE)

SENSORS

ARGOS	Platform positioning and data collection system
ATRS	Along Track Scanning Radiometer
AVHRR	Advanced Very High Resolution Radiometer
CZCS	Costal Zone Color Scanner
ESMR	Electronic Scanning Microwave Radiometer
LIDAR	Light Detection and Ranging
MSR	Microwave Scanning Radiometer
MSU	Microwave Sounding Unit
SASS	Seasat - A Satellite Scatterometer
SMMR	Scanning Multi-Channel Microwave Radiometer
SSMI	Special Sensor Microwave Imager
SODAR	Sound Detection and Ranging
VHRR	Very High Resolution Radiometer

LIST OF PARTICIPANTS

Bernstein, Robert L.
California Space Institute, A-021
Scripps Institution of Oceanography
University of California, San Diego
La Jolla, California 92093
USA

Bretherton, Francis P.
National Center for Atmospheric Research
P.O. Box 3000
Boulder, Colorado 80307
USA

Campbell, Garrett G.
CIRA
Colorado State University
Fort Collins, Colorado 80523
USA

Deschamps, Pierre-Yves
CNES
18, Avenue E. Belin
31055 Toulouse
FRANCE

Fellous, Jean-Louis
CNES - BPP - PR
129, rue de l'Université
75007 Paris
FRANCE

Fieux, Michele
Laboratoire d'Océanographie Physique
Museum d'Histoire Naturelle
43-45, rue Cuvier
75005 Paris
FRANCE

Gautier, Catherine
California Space Institute, A-021
Scripps Institution of Oceanography
University of California, San Diego
La Jolla, California 92093
USA

Gonella, Joseph
Laboratoire d'Océanographie Physique
Museum d'Histoire Naturelle
43-45, rue Cuvier
75005 Paris
FRANCE

Gower, James F.R.
Ocean Sciences
1230 Government Street
Victoria V8W 1YA
CANADA

Hojerslev, Niels
Institute of Physical Oceanography
University of Copenhagen
Haraldsgade 6
DK-2200 N- Copenhagen
DENMARK

Large, William G.
National Center for Atmospheric Research
P.O. Box 3000
Boulder, Colorado 80307
USA

Le Borgne, Pierre
Centre de Météorologie Spatiale
BP 147
22302 Lannion
FRANCE

Lefèbvre, Michel
GRGS-CNES
18, Avenue E. Belin
31055 Toulouse
FRANCE

Liu, Timothy W.
Jet Propulsion Laboratory
Mail Stop 183-501
Pasadena, California 91109
USA

Marsouin, Anne
Centre de Météorologie Spatiale
BP 147
22302 Lannion
FRANCE

McWilliams, James C.
National Center for Atmospheric Research
P.O. Box 3000
Boulder, Colorado 80307
USA

Minnet, Peter J.
Rutherford Appleton Laboratory
Chilton, Didcot
Oxfordshire, OX11 0QX
UNITED KINGDOM

Minster, Jean-François
Institut Physique du Globe
Université Paris VI
75005 Paris
FRANCE

Morrison, John
National Science Foundation
Room 611, 1800 G. Street, NW
Washington, D.C. 20550
USA

Périgaud, Claire
Institut Physique du Globe
Université Paris VI
75005 Paris
FRANCE

Salusti, Ettore
Instituto di Fisica
Roma
ITALY

Sarachick, Edward S.
Center for Earth & Planetary Physic
Harvard University
29 Oxford Street
Cambridge, Massachussetts 02138
USA

Taylor, Peter K.
Institute of Oceanographic Science
Brook Road
Wormley, Godalming
Surrey, GU8 5UB
UNITED KINGDOM

Vasconcellos, Michael de
Rua Fialho de Almeida, 1-4 Esq.
1000 Lisbon
PORTUGAL

Webster, Ferris
College of Marine Studies
University of Delaware
Lewes, Deleware 19958
USA

Wilson, Stanley W.
NASA Headquarters
Code EE-8
Washington, D.C. 20546
USA

Woods, John
Institut fur Meereskunde
and der Universitat Kiel
Dusternbrooker Weg 20
D-2300 Kiel
GERMANY

Wylie, Donald P.
Space Science and Engineering Center
University of Wisconsin
1225 W. Dayton Street
Madison, Wisconsin 53706
USA

INDEX